AF798931

NanoScience and Technology

Series Editors

Phaedon Avouris, IBM Research—Thomas J. Watson Research, Yorktown Heights, USA

Bharat Bhushan, Mechanical and Aerospace Engineering, The Ohio State University, Columbus, USA

Dieter Bimberg, Center of NanoPhotonics, Technical University of Berlin, Berlin, Germany

Cun-Zheng Ning, Department of Electronic Engineering, Tsinhua University, Beijing, China

Klaus von Klitzing, Max Planck Institute for Solid State Research, Stuttgart, Germany

Roland Wiesendanger, Department of Physics, University of Hamburg, Hamburg, Germany

The series NanoScience and Technology is focused on the fascinating nano-world, mesoscopic physics, analysis with atomic resolution, nano and quantum-effect devices, nanomechanics and atomic-scale processes. All the basic aspects and technology-oriented developments in this emerging discipline are covered by comprehensive and timely books. The series constitutes a survey of the relevant special topics, which are presented by leading experts in the field. These books will appeal to researchers, engineers, and advanced students.

Levente Máthé

Physical Properties of Low-Dimensional Systems

A Theoretical Study on Quantum Dots, Nanowires and Graphene

Springer

Levente Máthé
Materials
Energy and Advanced Technologies
National Institute for Research
and Development of Isotopic and Molecular
Technologies
Cluj-Napoca, Romania

ISSN 1434-4904 ISSN 2197-7127 (electronic)
NanoScience and Technology
ISBN 978-3-031-91601-4 ISBN 978-3-031-91602-1 (eBook)
https://doi.org/10.1007/978-3-031-91602-1

© The Editor(s) (if applicable) and The Author(s), under exclusive license to Springer Nature Switzerland AG 2025

This work is subject to copyright. All rights are solely and exclusively licensed by the Publisher, whether the whole or part of the material is concerned, specifically the rights of translation, reprinting, reuse of illustrations, recitation, broadcasting, reproduction on microfilms or in any other physical way, and transmission or information storage and retrieval, electronic adaptation, computer software, or by similar or dissimilar methodology now known or hereafter developed.
The use of general descriptive names, registered names, trademarks, service marks, etc. in this publication does not imply, even in the absence of a specific statement, that such names are exempt from the relevant protective laws and regulations and therefore free for general use.
The publisher, the authors and the editors are safe to assume that the advice and information in this book are believed to be true and accurate at the date of publication. Neither the publisher nor the authors or the editors give a warranty, expressed or implied, with respect to the material contained herein or for any errors or omissions that may have been made. The publisher remains neutral with regard to jurisdictional claims in published maps and institutional affiliations.

This Springer imprint is published by the registered company Springer Nature Switzerland AG
The registered company address is: Gewerbestrasse 11, 6330 Cham, Switzerland

If disposing of this product, please recycle the paper.

To my mom!

Preface

The progress of nanotechnology and nanofabrication has ushered a new era in electronics by enabling fabrication, measurement, and control at the atomic scale, where quantum processes play a significant role. In today's nanoelectronic devices, discrete energy levels and quantum states of matter are present, and phenomena such as quantum tunneling, interference, and many-body effects are exploited for novel functionalities.

Among the key devices of this new era in electronics are the quantum dot-based devices, used in both photonics (e.g., light-emitting diodes, detectors) and electronics (e.g., single-electron transistors, nonvolatile memories). Quantum dots are essentially artificial atoms, and have discrete levels of energy that can be populated with electrons, making them highly controllable and sensitive to the presence of nearby charges. This property has made quantum dots an ideal platform for detecting Majorana bound states in topological superconducting nanowires. The Majorana bound states are localized edge modes that obey non-Abelian statistics, providing nonlocal electronic degrees of freedom for encoding quantum information. This makes them a promising platform for topological quantum computation, as the nonlocal nature of Majorana qubits offers robust protection against decoherence, enabling the realization of decoherence-free quantum computation. Thus, the theoretical and experimental study of such systems is highly relevant. Quantum dots connected to leads, acting as electron reservoirs, offer a highly controllable environment for investigating many-body effects due to the precise control over parameters.

The current book focuses on the theoretical exploration of quantum transport processes in quantum dot-based nanodevices, as well as the analysis of the optical properties of spherical quantum dots. The book provides a coherent and detailed presentation of low-dimensional structures, starting with quantum dots (having 0-dimension), throughout nanowires (1-dimension) and graphene (2-dimensions).

In Chap. 1, we introduce fundamental concepts related to the topics, including quantum transport in low-dimensional systems, the Coulomb blockade, the Kondo effect, and the physics of Majorana fermions and Majorana bound states, presented accessibly. Additionally, we provide a brief review of the electronic properties of graphene and discuss the optical intersubband transition, and electric polarization, characterizing the optical properties of two-level quantum systems realized in semiconductor quantum dots.

In Chap. 2, we propose an experimentally feasible nanojunction design, where a quantum dot is integrated into a semiconductor-superconductor heterostructure hosting Majorana bound states in the presence of an optical phonon mode. A threading magnetic flux is employed to control Majorana-induced phonon-assisted quantum tunneling. This device can serve as a promising guide in Majorana detection experiments.

In Chap. 3, we present a detailed theoretical study of quantum transport in a system where a quantum dot is embedded between two graphene electrodes, under the influence of a magnetic field. This setup serves as an ideal platform for exploring many-body quantum phenomena, such as the Kondo effect, and offers valuable insights for advancing graphene-based nanoelectronic devices.

In Chap. 4, we theoretically study the linear and nonlinear optical properties of a spherical semiconductor quantum dot, where quantum confinement is approximated with an inversely quadratic Hellmann potential. The proposed setup could be further utilized to boost the development of novel quantum dot-based optoelectronic devices.

Cluj-Napoca, Romania Levente Máthé

Acknowledgments The present book is an adaptation of my Ph.D. thesis, defended on 15 September 2023, at the Doctoral School of Physics, Babeş-Bolyai University, Cluj-Napoca, Romania.

Firstly, I would like to thank my former academic supervisor, Prof. Dr. Ioan Grosu, for his endless patience, for always being there to answer my questions and for his constant scientific support during my Ph.D. studies. I am especially grateful to him for encouraging and motivating me to further my interest in the field of theoretical physics!

Special thanks are addressed to my Laboratory Chief at the National Institute for Research and Development of Isotopic and Molecular Technologies (INCDTIM), Dr. Liviu P. Zârbo, for welcoming me in the Quantum Engineering group, and for his continuous support throughout the creation of present book. Thank you, Liviu, for smoothing my scientific path!

I would like to express my honest gratitude to my colleague and friend from INCDTIM, Dr. Doru Sticleţ, for always being ready to help me and teach me, by coming up with a solution for every problem I encountered, whether in physics, mathematics or programming. Doru, thank you for your sincere patience with me!

Additionally, I would like to extend my sincere gratitude to Prof. Dr. Cătălin Paşcu Moca from the University of Oradea for his invaluable support in proofreading and for being readily available for insightful discussions on scientific and technical matters, which significantly contributed to the quality of this book.

I am very grateful to my friend Dr. Sándor Borbély, the Vice Dean of Faculty of Physics, Babeş-Bolyai University for always being available to answer my questions and for providing valuable feedback.

Last but not least, I am deeply grateful to my beloved parents, Pál and Margit, my brother, Attila, and my love, Imola, for always being there for me, for their endless moral support and the immense patience they have shown me while writing the present book.

I sincerely thank the Springer editorial staff for their invaluable assistance throughout the book's publication process.

SDG!

This book was supported through the "Nucleu" Program within the National Research Development and Innovation Plan 2022–2027, Romania, carried out with the support of MEC, project no. 27N/03.01.2023, component project code PN 23 24 01 04.

Cluj-Napoca, Romania Levente Máthé

Competing Interests The author has no competing interests to declare that are relevant to the content of this manuscript.

Contents

Abbreviations

CAR	Crossed Andreev Reflection
DFT	Density Functional Theory
DOS	Density of States
EOM	Equation of Motion
EPI	Electron-Phonon Interaction
ET	Electron Tunneling
IQH	Inversely Quadratic Hellmann
LAR	Local Andreev Reflection
MBS	Majorana Bound State
MF	Majorana Fermion
MZM	Majorana Zero Mode
NCA	Noncrossing Approximation
NGF	Nonequilibrium Green's Function
NRG	Numerical Renormalization Group
NU	Nikiforov-Uvarov
QD	Quantum Dot
QMC	Quantum Monte Carlo
SBMF	Slave-Boson Mean-Field
SC	Superconductor
STM	Scanning Tunneling Microscopy
TSNW	Topological Superconducting Nanowire

Chapter 1
Theoretical Background

This chapter of the book serves as an introductory section, focusing on a brief discussion of basic concepts related to the book's subject matter. It includes a concise overview of quantum transport in low-dimensional systems, such as quantum dots, by presenting the primary phenomena that occur in such semiconductor heterostructures, namely the Coulomb blockade and Kondo effect. Next, this section provides insights into the complex physics of Majorana fermions and Majorana bound states by discussing the fundamental theoretical and experimental efforts in this area. The main models serving to the creation of Majorana bound states in superconductor-based systems, specifically the Kitaev and Oreg-Lutchyn models, are presented. Additionally, the electronic properties of graphene are reviewed. The final part of this chapter further explores the main phenomena, the optical intersubband transition and electric polarization, characterizing the optical properties of two-level quantum systems realized in semiconductor quantum dots. The linear and third-order nonlinear susceptibilities, treated within the density-matrix approach using perturbation theory, are discussed.

1.1 Introduction to the Quantum Transport in Low-Dimensional Systems

Advances in nanotechnology have made it possible to miniaturize electronic components of controlled geometries at atomic scale where quantum phenomena occur naturally. Such semiconductor structures are ideal platforms for studying the quantum transport of charge carriers under adjustable conditions requiring the necessity of theoretical investigations [1, 2]. As we know from classical physics, the conductance of a macroscopic conducting wire decreases as its cross-section decreases and increases with the decrease of its length. This scaling law (called ohmic law) is not valid for conductors of dimensions comparable with the phase-coherence length of

© The Author(s), under exclusive license to Springer Nature Switzerland AG 2025
L. Máthé, *Physical Properties of Low-Dimensional Systems*,
NanoScience and Technology, https://doi.org/10.1007/978-3-031-91602-1_1

electrons where quantum nature plays an important role [3–5]. The phase-coherence length means the distance that the electron travels without changing the phase of its associated wavefunction. By definition, a system is considered mesoscopic if the characteristic size of the transport region is smaller than the electron phase-coherence length, but it is much larger than its de Broglie wavelength. However, a system is nanoscale if the characteristic size of the transport region is comparable with the de Broglie wavelength of the electron and exhibits discrete energy spectrum. A nano-junction is realized by embedding a *confined region* (e.g., semiconductor quantum dots (QDs) or single molecules) between electron reservoirs formed from macroscopic electrodes [6, 7]. The confined region couples to the leads via tunnel barrier as it is shown in Fig. 1.1a. The transport takes place between the leads through the confined region. In the absence of interactions (e.g., electron-electron, electron-vibron), the quantum coherent transport through the junction can be determined within the Landauer-Büttiker formalism, in which the phase-coherence of electrons is preserved during their motion across the junction. The Landauer-Büttiker method allows the calculation of current and conductance in the linear voltage regime based on the wavefunction of electrons in junction. In the presence of strong interactions and at finite voltages, the Landauer-Büttiker method breaks down, and the use of other techniques, such as the quantum master equation or nonequilibrium Green's function (NGF) methods, is required [7].

1.1.1 Quantum Dots. The Coulomb Blockade

In the present book, we employ the NGF technique to determine the transport properties of nanojunctions consisting of QDs with electron-phonon (see Chap. 2) or electron-electron (see Chap. 3) interactions. The NGF method allows for the study of the Coulomb blockade and Kondo effect in nanojunctions such as QDs. QDs are small confined regions of dimensions comparable with the electron de Broglie wavelength, considered as zero-dimensional objects, are embedded between semi-infinite electrodes which play the role of electron reservoirs. The dot tunnel couples to the lead left (right) with coupling $\Gamma_{L(R)}$. The motion of electrons is restricted in all spatial directions, and thus they occupy discrete energy levels in the QD, which can be probed in transport measurements via macroscopic electrodes. The capacitively coupled gate electrode allows the fine-tuning of energy levels and modifying the electron number in QD. A typical QD setup is illustrated in Fig. 1.1a with its discrete energy levels in Fig. 1.1b. In the case of an uncoupled QD ($\Gamma_L = \Gamma_R = 0$), the total charge is quantized as $Q = -Ne$ with N and e being the number of electrons in the confined region and the elementary charge. When the dot couples to the electrodes, to put an extra electron in the QD or remove it from the QD, the Coulomb charging energy, i.e., $U = e^2/2C$ (where C represents the capacitance of the QD), should be overcome. The condition that the dot exhibits quantized energy levels is first that the Coulomb energy has to be much larger than the thermal excitation energy ($k_B T$), i.e., $k_B T \ll U$. This condition requires that the capacitance of the dot to be very

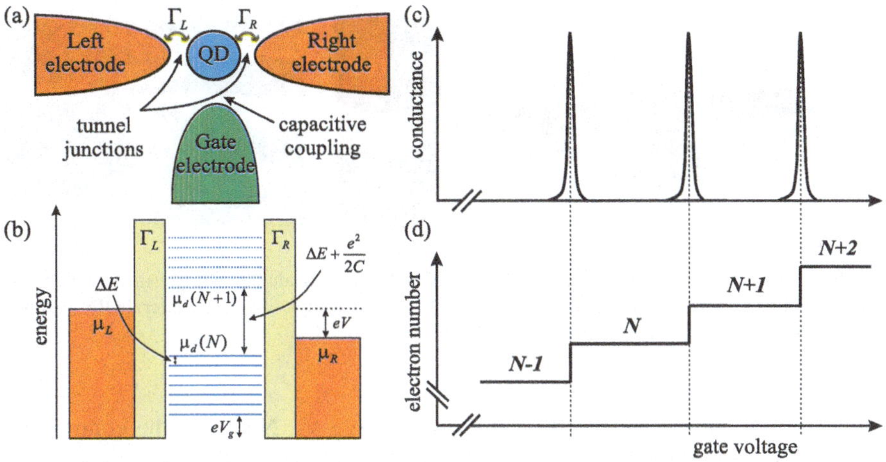

Fig. 1.1 **a** Representation of a quantum dot (QD) embedded between two macroscopic left (source) and right (drain) electrodes playing the role of electron reservoirs. The QD couples to the leads via tunnel junctions with couplings Γ_L and Γ_R, respectively. The gate electrode is capacitively coupled to the QD. Adapted with permission from [1, 8]. Copyrights 1997, Springer Nature. Copyright 2002 by the author. **b** Energy diagram of a QD which levels can be tuned by gate voltage (V_g) applied to the gate electrode. The notations $\mu_d(N)$, ΔE and $e^2/2C$ refer to the electrochemical potential in the dot, to the level spacing and to the Coulomb charging energy with C and e being the QD capacitance and elementary charge, respectively. The reservoirs have continuum energy spectrum, where the chemical potential $\mu_{L(R)}$ in lead left (right) defines the bias voltage $V = (\mu_L - \mu_R)/e$. The transport is blocked, while the QD contains N number of electrons. This situation corresponds to the Coulomb blockade. Adapted with permission from [1, 2, 9]. Copyrights 1997, 1999, Springer Nature. Copyright 2010, Taylor & Francis Group LLC-Books. **c** The conductance as a function of gate voltage V_g. Adapted with permission from [1, 2]. Copyrights 1997, 1999, Springer Nature. **d** The variation of electron number in the dot with the gate voltage V_g. Adapted with permission from [1, 2]. Copyrights 1997, 1999, Springer Nature

small. Second, the QD should be weakly connected to the electrodes, in order the electron to be situated on the dot or in the electron reservoirs, i.e., $\Gamma_{L(R)} \ll U$. The last condition requires a high dot-lead tunneling resistance (R_t), which, based on Heisenberg's uncertainty principle, is $R_t \gg h/e^2$ [1, 2, 6, 8, 10]. In addition, the spatial confinement of the motion of electrons leads to a quantum mechanical quantization in the spectrum of the QD. The confinement induced energy levels can be controlled via adjusting the potential characteristics. The total QD spectrum is a result of the confinement induced and the charge quantization generated energy spectrum [1, 2, 8]. Figure 1.1b shows the discrete energy spectrum of a QD with confined induced quantum mechanical energy states with splitting, ΔE, and the Coulomb interaction generated many-body contribution, $U = e^2/2C$.

Let us consider that the dot states are filled with N electrons up to the level $\mu_d(N)$. The lowest $(N + 1)th$ empty state corresponds to $\mu_d(N + 1)$ [see Fig. 1.1b]. At a given gate voltage V_g, to add an extra electron in the QD, one needs to overcome the energy gap $\mu_d(N + 1) - \mu_d(N) = \Delta E + e^2/2C$. The electrochemical potential

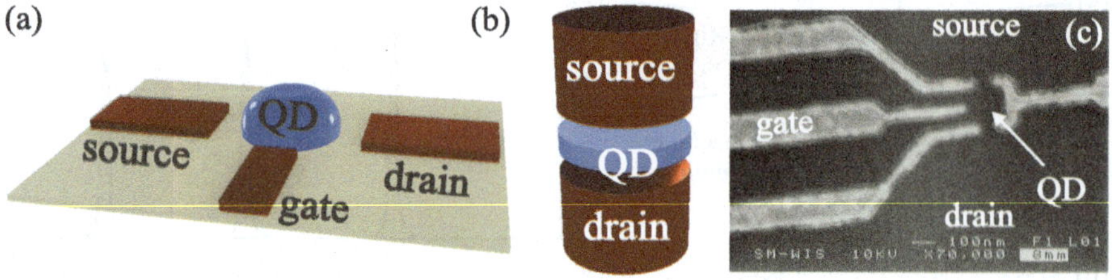

Fig. 1.2 Representation of a **a** lateral and **b** vertical QD. Figure **b** adapted with permission from [1]. Copyright 1997, Springer Nature. **c** Scanning Electron Microscopy image of a lateral QD. Adapted with permission from [12]. Copyright 1998, Springer Nature

$\mu_d(N)$ in the dot can be tuned by the gate voltage V_g. Note that the contribution $U = e^2/2C$ from the Coulomb interaction exists only at $\mu_d(N)$. Below the energy $\mu_d(N)$ there are only the quantum states with energy spacing ΔE [1, 2]. The dot contains N localized electrons. Due to the fact that the electronic state $\mu_d(N)$ is occupied, no electron from the leads can enter the dot, because the next lowest empty state $\mu_d(N+1)$ is located above the chemical potentials in the electrodes. In this case, the electron number in the QD is fixed and the transport between the electrodes is forbidden. This phenomenon is called *Coulomb blockade*. In order to realize a current flow through the dot, the $(N+1)th$ level has to be located in the bias voltage window defined by the leads' chemical potentials, $V = (\mu_L - \mu_R)/e$. When tuning the dot level $\mu_d(N+1)$ within the energy window $\mu_R < \mu_d(N+1) < \mu_L$, then an electron from left lead can tunnel to the QD and can leave it to the right lead opening thus a transport channel from left lead to right one. In this case, the electron number can fluctuate between N and $(N+1)$ [1, 2, 8]. Similarly, if the state $\mu_d(N+1)$ is filled and by tuning the next empty level $\mu_d(N+2)$ in the voltage window $V = (\mu_L - \mu_R)/e$, an additional transport channel will be opened, resulting Coulomb oscillations as a function of gate voltage. This situation is depicted in Fig. 1.1c where the linear conductance is plotted as a function of applied gate voltage V_g. The variation of electron number with gate voltage is illustrated in Fig. 1.1d.

According to geometrical arrangement of QDs, we can distinguish lateral [Fig. 1.2a] and vertical [Fig. 1.2b] QDs. Figure 1.2c illustrates an experimental realization of a lateral QD. The first QDs were mostly fabricated in GaAs semiconductor heterostructures [11–15].

1.1.2 *The Kondo Effect in Quantum Dots*

As known from solid states physics, the electrical resistance of nonmagnetic metals (e.g., Cu, Au) decreases when the temperature is lowered resulting in a reduction of the lattice vibrations and the electrons can more easily pass the metallic crystal. A further decrease in temperature leads to the saturation of resistivity which is related to the electron scattering on the static defects in the crystal lattice [10, 16]. By cleaning the sample, i.e., reducing the number of defects in the material, the saturation resistance can be decreased. A group of metals (e.g., Al, Nb, Pb) may abruptly lose all their electrical resistance at a critical temperature when undergoing a phase transition from the conducting to a superconducting state. It has been observed that when magnetic atoms (e.g., Mn, Cr, Fe,) are added to nonmagnetic metals, the resistance-temperature behavior of material is drastically changed. Namely, the electrical resistance starts to increase as the temperature further decreases, without exhibiting a phase transition. The temperature at which the electrical resistivity begins to grow again is called the Kondo temperature. The last one determines the electronic properties of materials at low temperatures [10, 16, 17]. Such kind of resistance-temperature behavior in Au samples was observed in 1934 [18], but for an explanation had to wait until 1964. Namely, Kondo, by considering the interaction of magnetic impurities with a sea of conduction electrons, studied the scattering from magnetic ions by employing perturbation theory [19]. It was found that scattering from magnetic ions leads to a logarithmic grow in resistivity when lowering the temperature. It turned out that the Kondo theory accurately describes the behavior of resistivity in the low-temperature regime, above the Kondo temperature. However, it gives unphysical results below the Kondo temperature, when the resistivity diverges to infinite as the temperature tends to zero [16]. In the following years, a major effort has been paid to determine the dependence of resistance on temperature well below the Kondo temperature range [20, 21].

As mentioned previously, the developing of Kondo effect is a result of interaction of spin-dependent electrons with a localized magnetic impurity. Let us consider a single-level spin-degenerate QD, for which, N can be equal to 0, 1 or 2. The model is known as the Anderson impurity model, introduced in 1961 for the study of a single magnetic impurity embedded in nonmagnetic metal host [22]. We assume low-temperature regime where charge fluctuations in dot are suppressed, i.e., $k_B T \ll |\varepsilon_d|$. We first consider the case when the dot level ε_d is filled by two electrons with opposite spin direction ($N = 2$), the first-order tunneling process (discussed above) is blocked because the next empty level at $\varepsilon_d + U$ is located above the leads' chemical potential $\mu_{L(R)}$ [see Fig. 1.3a]. Note that the total spin of the QD is zero. Due to the Heisenberg's uncertainty principle, an electron can escape the dot leading to an unstable one-electron intermediate (virtual) state in the QD existing for a small time $\tau \sim h/|\varepsilon_d|$. Considering that the escaped electron enters the right lead and it is replaced by other one from the left lead, one electron transfer has been realized between the electrodes. This phenomenon is called second-order elastic cotunneling because the energies of the initial state and the final state are the same. The situation is different for odd numbers of electrons on dot level $N = 1$ when the QD has an $1/2$-spin [see Fig. 1.3b]. Similarly to the previous case, an electron with spin-up can tunnel out the dot and enter an unoccupied state in the right lead. The escaped electron can be replaced by another one with opposite spin direction coming from the left electrode, leading to a flip of the QD spin from spin-up (-down) to spin-down (-up). This type of transport is called second-order spin-flip cotunneling, in which the initial and final states of QD have different spin orientations. The second-order spin-flip cotunneling process

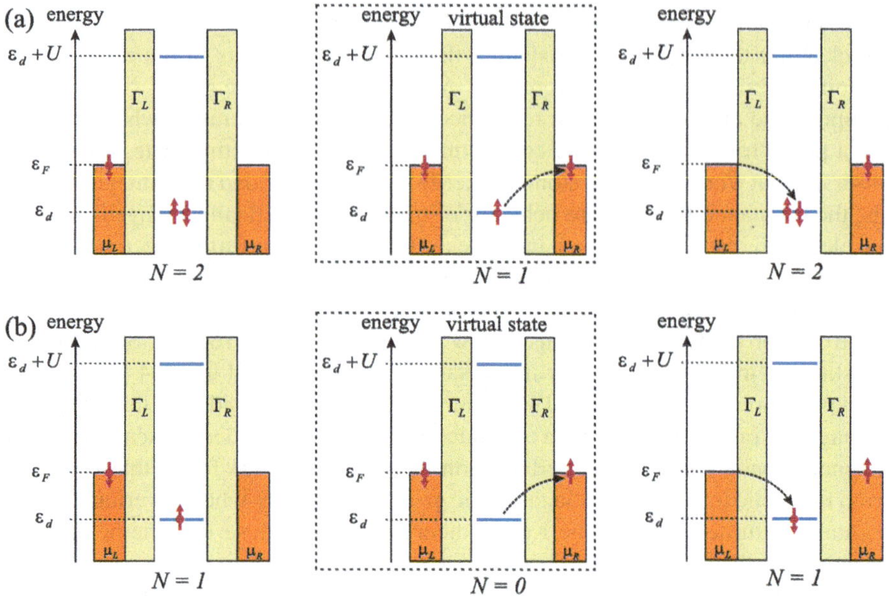

Fig. 1.3 **a** Representation of the second-order elastic cotunneling process in a QD, coupled to leads via couplings $\Gamma_{L(R)}$, having two electrons ($N = 2$) and zero-spin. Adapted with permission from [9]. Copyright 2010, Taylor & Francis Group LLC-Books. **b** Representation of the second-order spin-flip cotunneling process in QD with one electron and 1/2-spin. The notations are: ε_d-dot energy, U-Coulomb energy and $\mu_{L(R)}$-chemical potential in lead left (right) is equal to the Fermi level ε_F. For explanation see the text. Adapted with permission from [8, 9, 13]. Copyright 2010, Taylor & Francis Group LLC-Books. Copyright 2002 by the author. Copyright 1998, The American Association for the Advancement of Science

realizes an antiferromagnetic exchange between the spin of leads' electrons and the QD spin. This results in an entire screening of the localized QD spin by the spin cloud of conduction electrons at temperatures lower than a characteristic value, known as the Kondo temperature T_K, which represents a characteristic energy scale. The antiferromagnetic exchange leads to the formation of a many-body singlet between the electrons of leads and the localized dot electron, opening thus an additional transport channel at the Fermi energy by forming a resonance [9, 16, 23]. The peak emerging at the Fermi level in the density of states is called Kondo resonance [see Fig. 1.4a]. In nonequilibrium ($\mu_L \neq \mu_R$), two Kondo resonances with smaller amplitude develop in the QD density of states at the chemical potentials in the leads [Fig. 1.4b]. At temperatures lower than the Kondo temperature, the differential conductance exhibits a narrow peak at zero-bias voltage, referred to as the zero-bias peak (or zero-bias anomaly) or Kondo (or Abrikosov-Suhl) resonance, which mimics the Kondo resonance in the density of states [9, 13, 15, 23]. Contrary to bulk metals where the Kondo effect emerges as a ln T divergence in resistance [Fig. 1.4c], in QDs it shows up as a ln T dependence of the linear conductance [Fig. 1.4d]. The reason of the difference in developing of Kondo effect in bulk metals and QDs can be

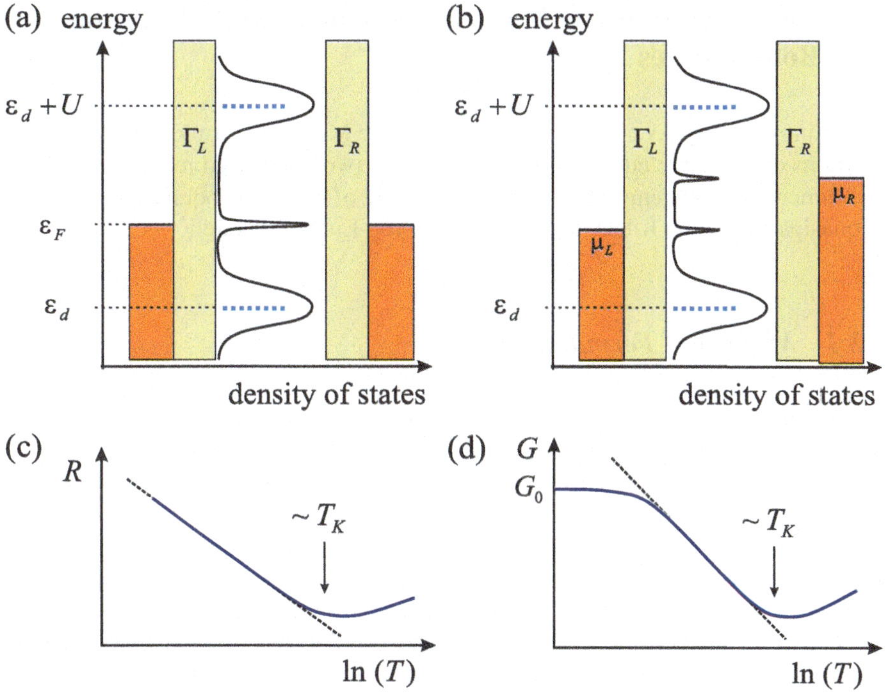

Fig. 1.4 Schematic illustration of the density of states of a QD, coupled to leads via couplings $\Gamma_{L(R)}$, at **a** equilibrium $\mu_L = \mu_R = \varepsilon_F$ and **b** out of equilibrium $\mu_L \neq \mu_R$. Here, ε_F is the Fermi energy in leads and $\mu_{L(R)}$ is the chemical potential in lead left (right). Adapted with permission from [8, 13]. Copyright 2002 by the author. Copyright 1998, The American Association for the Advancement of Science. **c** The logarithmic dependence of resistance R on temperature T in bulk metals. Adapted with permission from [9]. Copyright 2010, Taylor & Francis Group LLC-Books. **d** The logarithmic dependence of conductance G on temperature T in QDs where $G_0 = 2e^2/h$ is the conductance quantum. Adapted with permission from [9]. Copyright 2010, Taylor & Francis Group LLC-Books

explained with their geometries [9, 16]. In a bulk metal, the spin-flip processes due to the scattering of electrons on magnetic impurities results in a mixing of electron wavefunctions with different momenta which leads to an increased scattering cross-section and thus to an enhanced resistance. In QD systems where the dot couples to electron reservoirs, the electrons can only pass through the dot, without bypassing it. Therefore, the spin-flip cotunneling processes can lead to the mixing of electron wavefunctions from different electrodes by increasing thus the conductance through the dot [9, 16]. In the unitary limit, the zero-temperature linear conductance reaches the $2e^2/h$ value with symmetrical couplings $\Gamma_L = \Gamma_R$. The Kondo effect was first experimentally demonstrated in GaAs-based QDs [12–15].

1.2 Introduction to the Majorana Fermions and Majorana Bound States

In this introductory section, we briefly review the Majorana fermions (MFs) as real solutions of the Dirac equation. We next present two condensed matter concepts of one-dimensional systems which lead to emerging of Majorana bound states (MBSs) as quasiparticles: the Kitaev toy model and Oreg-Lutchyn model.

1.2.1 Majorana Fermions

In the early part of the last century, we have witnessed quite a lot of efforts to develop the present knowledge regarding the quantum mechanics throughout history. Namely, Heisenberg created the matrix mechanics in 1925 as a first formulation of quantum mechanics. Soon thereafter, Schrödinger presented his non-relativistic wave equation which describes the quantum motion of fundamental particles. Shortly afterwards, Dirac published his wave equation which describes the quantum behavior of relativistic particles which led to the prediction of antimatter [24]. Mathematically, the antiparticle emerges as a second complex solution beside an existing one of the Dirac equation . The antiparticle has negative energy with opposite charge and same spin with respect to the particle [25]. Majorana recognized in 1937 that the Dirac equation can be represented in terms of real wavefunctions [26]. The representation entails an important consequence, i.e., a particle described by a real wavefunction is its own antiparticle. According to Majorana's prediction, *neutrinos* could be a candidate for such relativistic fermions which is still remained an open problem in the particle physics [27]. Note that the first experimentally detected antiparticle was the *positron*, in 1932 [28]. Unfortunately, the detection of such fermions that fulfill the Majorana equation, is still to be done, remaining thus an experimental challenge for physicists working in the area of elementary particle physics, nuclear physics and cosmology [24].

Let us consider the three-dimensional Dirac equation for a particle with spin-1/2 and mass m in the presence of an electromagnetic field $\mathcal{A}_\mu$ as

$$(i\Gamma^\mu(\partial_\mu - ie\mathcal{A}_\mu) - m)\psi = 0, \tag{1.1}$$

with notation $\partial_\mu \equiv \partial/\partial x^\mu$ where the Γ^μs are 4x4 Dirac matrices and obey the Clifford's algebra, i.e., $\{\Gamma^\mu, \Gamma^\nu\} = 2\eta_{\mu\nu}$ where $\eta_{\mu\nu}$ is the Minkowski metric tensor (and $\hbar = c = 1$) [24, 29]. We now define the antiparticle solution, ψ_c, to the Dirac equation (1.1) as a result of the charge conjugation operation. The action of charge conjugation operator $\mathcal{C}$ on Dirac equation leads to

$$\mathcal{C}[(i\Gamma^\mu(\partial_\mu - ie\mathcal{A}_\mu) - m)\psi] = (i\Gamma^\mu(\partial_\mu + ie\mathcal{A}_\mu) - m)\mathcal{C}\psi = 0, \tag{1.2}$$

where the charge conjugate solution is $\psi_c = \mathcal{C}\psi$. The condition that a particle to be equal to its own antiparticle requires $\psi_c = \mathcal{C}\psi = \psi$ which is Lorentz invariant. This yields that only neutral ($e = 0$) spin-$1/2$ fermions satisfy the relation $\psi_c = \psi$. Because of the charge conjugation operator $\mathcal{C}$ is anti-unitary, it involves a complex conjugation operation [24, 29]. Consequently, the condition for existing MFs is

$$\psi^* = \psi. \tag{1.3}$$

1.2.2 Majorana Bound States

The detection of MFs has come under the spotlight in the last decade in condensed matter community [30]. As a matter of fact, the Majorana equation possesses the same mathematical structure as the Bogoliubov-de Gennes equation. The last one describes the quasiparticle excitations in superconductors (SCs) for which there is a particle-hole (charge conjugation) symmetry. The quasiparticles in SCs imply coherent superposition of electrons and holes. The electrons correspond to filled states with energy above the Fermi level, whereas the holes correspond to empty states with energy below the Fermi level and opposite charge. Consequently, the excitations at the Fermi level, in the middle of superconducting gap, are charge-neutral and they are a result of coherent superposition of electrons and holes, which stand for the Majorana fermions [24, 31].

Let us consider the annihilation (creation) Majorana operator γ ($\gamma^\dagger$). The following relation has to be satisfied:

$$\gamma = \gamma^\dagger, \tag{1.4}$$

showing that the operators are Hermitian. The Majorana operator can be thus expressed as a sum of regular fermionic operators associated to a particle and its antiparticle,

$$\gamma = f + f^\dagger, \tag{1.5}$$

where $f^\dagger$ (f) is the creation (annihilation) operator for a regular fermion. As mentioned previously, due to the similarity of Bogoliubov-de Gennes and Majorana equations, the Majorana excitations can be searched in SCs. Such quasiparticle excitations within the BCS theory can be described as a linear combination of annihilation and creation operators,

$$\gamma_\sigma(E) = u_{\sigma,E} f_\sigma + v_{-\sigma,E} f^\dagger_{-\sigma}, \tag{1.6}$$

where $u_{\sigma,E}$ and $v_{-\sigma,E}$ are energy (E) and spin ($\sigma =\uparrow$ and $-\sigma =\downarrow$) dependent amplitudes. A Majorana excitation requires equal amplitudes of operators f_σ and $f^\dagger_{-\sigma}$. In addition, the spin index destroys the Majorana condition, because $\gamma_\sigma(E) \neq \gamma^\dagger_\sigma(-E)$. These conditions are fulfilled for midgap excitations with zero excitation energy

($E = 0$) for spinless fermions in SCs, i.e., $u_{E=0} = v^*_{E=0}$. Consequently, Majoranas are zero-energy excitations in SCs, for which the terms Majorana zero modes (MZMs) or equivalently Majorana bound states (MBSs) are used in the literature [24, 25, 29, 32]. The MFs in condensed matter can be created as zero-energy excitations in spinless or p-wave SCs with spin-triplet pairing [24, 29].

As seen previously, MFs are charge-neutral particles corresponding to real solutions of the Dirac equation and thus they are their own antiparticles. The MBSs appear as a pair of spatially separated localized edge modes in topological SCs which can store information as a nonlocal regular fermionic state. The existence of MBSs is not influenced by local perturbations (e.g., disorder) and, in general, they are protected from quantum decoherence. The MBSs obey non-Abelian statistics which means that the exchange of two (or more) MBSs in real space is non-commutative. Thus, the final results of the operation depends on the order of exchanges, an operation called braiding of Majorana modes. The non-Abelian braiding property of MBS could provide an ideal platform for topological quantum computation by realizing decoherence-free Majorana qubits [24, 31, 33–38].

In the following, we discuss two models which allow to create MBSs in topological SCs. The first one is the Kitaev model involving spinless fermions in a one-dimensional p-wave SC [33]. The second one is the Oreg-Lutchyn model, which ensures a more realistic setup based on a semiconducting nanowire with strong spin-orbit coupling (e.g. InAs or InSb) located in proximity of a one-dimensional s-wave SC threaded by a magnetic field [39, 40]. The fingerprints of MBSs were first detected in transport measurements in such heterostructures as a zero-bias anomaly in differential conductance map [41–45]. Additional theoretical and experimental alternative implementations have been considered to create MZMs, such as, devices based on two-dimensional p-wave SCs [46, 47], or s-wave SCs with topological insulators [48], chains of magnetic impurities deposited on surface of SCs [49–56], half-metallic ferromagnets with SC [57, 58], magnetic nanoparticles on SCs [59], nanomagnets [60], cold atom wires [61], vortices in SCs [62–65], PbTe-Pb hybrid nanowires [66]. The list of Majorana realizations is growing from year to year. Note that the zero-bias conductance peak can also be attributed to other phenomena, such as the Kondo effect [67, 68], or it may indicate the presence of quasi-Majoranas [69, 70] or low-energy Andreev bound states [71–73] arising from disorder [70, 74–79] or smooth confinement potential [69–71, 77, 79–82], all of which complicate the detection of MBSs. Thus, the experimental identification of Majoranas in condensed matter systems still remains an active research area. Systems hosting MBSs show a series of interesting phenomena such as the fractional Josephson effect [83, 84], and Andreev reflection [85–89].

The Kitaev Model

We now discuss the first model introduced by Kitaev [33] to create spatially separated MBSs in a one-dimensional topological SC. The tight-binding Hamiltonian of spinless electrons on a p-wave superconducting lattice with N finite sites, illustrated in Fig. 1.5a, is

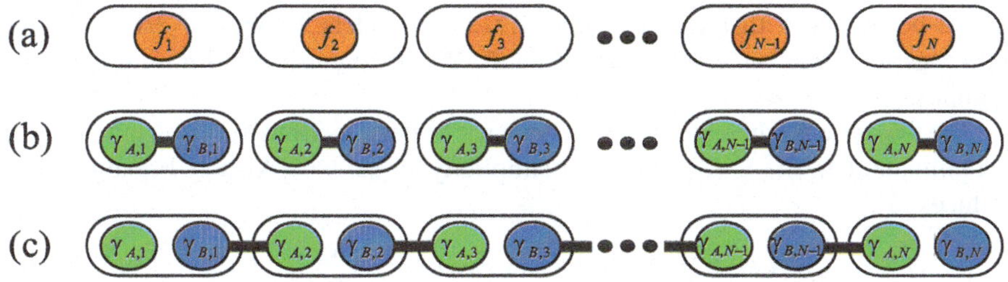

Fig. 1.5 **a** Representation of a one-dimensional spinless fermion chain denoted by regular fermion operators f_n. **b** Trivial phase of the chain: regular fermions are formed by paired MFs from the same physical lattice site represented by Majorana operators $\gamma_{A,n}$ and $\gamma_{B,n}$, respectively. **c** Topological phase: MFs from the same lattice site are decoupled and MFs localized on neighboring lattice sites are coupled resulting in the appearance of two unpaired Majorana modes at opposite ends of the chain. Adapted from [38] under CC-BY 4.0 license. Copyright 2022 by the author

$$\mathcal{H} = -\sum_{n=1}^{N} \mu f_n^\dagger f_n - \sum_{n=1}^{N-1} (t f_n^\dagger f_{n+1} + t^* f_{n+1}^\dagger f_n) + \sum_{n=1}^{N-1} (\Delta f_n f_{n+1} + \Delta^* f_{n+1}^\dagger f_n^\dagger), \tag{1.7}$$

where $f_n^\dagger$ (f_n) is the creation (annihilation) operator for a regular fermion on lattice site n. The first term in Hamiltonian (1.7) gives the fermion occupation on physical sites determined by the onsite energy μ (chemical potential). Note that the same lattice site is occupied by one spinless electron due to the Pauli exclusion. The second term in Eq. (1.7) models the interaction between neighboring lattice sites by hopping strength $t = t^* > 0$. The last part of $\mathcal{H}$ represents the creation of Cooper pairs where $\Delta = |\Delta| e^{i\varphi}$ stands for the superconducting pairing potential with φ being the superconducting phase. Note that φ can be absorbed by the unitary transformation $f_n \to f_n e^{-i\varphi/2}$ [25, 32, 38, 90]. Without loss of generality, for simplicity, we assume here that $\varphi = 0$. We now rewrite the Hamiltonian (1.7) of N complex fermions in terms of $2N$ Majorana operators. Mathematically, any complex fermion operator, described here by f_n, can be decomposed into two Majorana operators, denoted by $\gamma_{A,n}$ and $\gamma_{B,n}$, as follows:

$$f_n = \frac{1}{2}(\gamma_{A,n} + i\gamma_{B,n}), \tag{1.8}$$

with the property of Majorana operators $\gamma_{\nu,n} = \gamma_{\nu,n}^\dagger$, $\gamma_{\nu,n}^2 = 1$ and $\{\gamma_{\nu,n}, \gamma_{\beta,m}\} = 2\delta_{\nu\beta}\delta_{nm}$. Thus, the Hamiltonian (1.7) takes the form

$$\mathcal{H} = -\frac{1}{2}\sum_{n=1}^{N} \mu(1 + i\gamma_{A,n}\gamma_{B,n}) + \frac{1}{2}\sum_{n=1}^{N-1} [i(|\Delta| + t)\gamma_{B,n}\gamma_{A,n+1} + i(|\Delta| - t)\gamma_{A,n}\gamma_{B,n+1}]. \tag{1.9}$$

It can be observed that Hamiltonian (1.9) now models a chain of $2N$ Majorana fermions localized on sublattices A and B. We distinguish two limiting coupling regimes.

The first case is when the coupling between Majoranas from the same lattice site n is dominant, i.e., $|\mu| > t,\ |\Delta|$ and for simplicity we set $t = |\Delta| = 0$. Equation (1.9) reduces to

$$\mathcal{H} = -\frac{\mu}{2}\sum_{n=1}^{N}(1 + i\gamma_{A,n}\gamma_{B,n}) = -\mu\sum_{n=1}^{N} f_n^{\dagger} f_n, \tag{1.10}$$

which Hamiltonian expresses the localized regular fermions formed by paired MFs from the same physical lattice site, as it is shown in Fig. 1.5b. This phase is topologically trivial, which characterizes an atomic insulator [29].

The second case is when the coupling between MFs localized on neighboring lattice sites is stronger than the coupling strength between the Majoranas from the same lattice sites, i.e., $t,\ |\Delta| > |\mu|$. Thus, setting $|\mu| = 0$, and assuming that the superconducting gap is equal to the hopping strength, i.e., $|\Delta| = t$, the Hamiltonian in Eq. (1.9) becomes

$$\mathcal{H} = i\,t\sum_{n=1}^{N-1}\gamma_{B,n}\gamma_{A,n+1}. \tag{1.11}$$

Introducing a fermionic operator d_n as a superposition of Majorana operators from physical neighboring sites [38]

$$d_n = \frac{1}{2}(\gamma_{B,n} + i\gamma_{A,n+1}), \tag{1.12}$$

with $n = 1, 2, \ldots, N-1$, where the fermion anticommutation relations $\{d_n, d_m\} = 0$ and $\{d_n, d_m^{\dagger}\} = \delta_{nm}$ fulfill. In this case, Eq. (1.11) transforms into

$$\mathcal{H} = 2t\sum_{n=1}^{N-1}\left(d_n^{\dagger}d_n - \frac{1}{2}\right). \tag{1.13}$$

This results in the Majorana operators from the same lattice sites are now decoupled. We observe that the initial Hamiltonian (1.7) describes N fermions by operators f_n, while now the Hamiltonian (1.13) models $N-1$ fermions by operators d_n. The missing fermionic mode of operator d_N in Eq. (1.13) is formed by the missing Majorana modes of operators $\gamma_{A,1}$ and $\gamma_{B,N}$. These two unpaired Majorana fermions are completely localized at the edges of the lattice and are decoupled from the Hamiltonian (1.13). Since they commute with the Hamiltonian, $[\mathcal{H}, \gamma_{A,1}] = 0$, and $[\mathcal{H}, \gamma_{B,N}] = 0$, thus the missing fermionic mode can be expressed as the superposition of the two Majorana operators,

$$d_N = \frac{1}{2}(\gamma_{A,1} + i\gamma_{B,N}), \tag{1.14}$$

for which the fermion anticommutation relations $\{d_N, d_N\} = 0$ and $\{d_N, d_N^\dagger\} = 1$ are satisfied. This single fermion mode is delocalized between the two ends of the chain and is a result of the superposition of two Majorana modes localized at the opposite ends of the lattice, as it is illustrated in Fig. 1.5c. These Majorana end modes are the Majorana bound states (MBSs). Since this delocalized fermion mode does not appear in Hamiltonian (1.13), thus its energy level is fixed at zero. This phase is called topologically nontrivial [38].

Note that completely spatially separated MBSs can be realized only in systems with infinite-size. Contrary, real systems have finite sizes. Considering a finite system with size L, the two MZMs are located at the opposite edges of the topological SC at distance L from each other and their corresponding wavefunctions will overlap leading to a hybridization of MZMs. For a finite, non-zero μ, the MBSs are not perfectly localized at the edges of the chain and thus their wavefunctions exponentially decay with distance. In addition, the hybridization of MBSs results in a delocalized fermionic level with a finite energy ε_M. This ε_M energy is usually called Majorana overlap energy, and due to the exponentially decaying Majorana wavefunctions, ε_M decreases exponentially with distance L, i.e., $\varepsilon_M \propto e^{-L/\xi}$ where ξ is the superconducting coherence length, also called Majorana localization length [38]. For a long wire ($L \to \infty$), the MBSs are perfectly degenerate ($\varepsilon_M \to 0$).

The Oreg-Lutchyn Model: Majorana Nanowire

As seen in above, the Kitaev model is a spinless system involving the creation of Cooper pairs in a p-wave SC. Due to the fact that electrons have spin, the experimental realization of a spinless system can be feasible by energetic separation of the different spin orientations by applying a magnetic field, creating thus spin-polarization. Since p-wave SCs with spin-triplet pairing form the category of unconventional SCs, they appear rarely in the nature, and present low critical temperatures ($T < 1\,K$) with small superconducting gaps, and thus they should be replaced by other realizations in experiments [25, 38]. In addition, the superconductivity appearing in one-dimensional systems is not stable against quantum fluctuations at finite temperatures.

It has been established that a possible experimental realization of the Kitaev one-dimensional lattice chain model is obtained when depositing a semiconducting nanowire with strong spin-orbit coupling (e.g., InAs, InSb), as shown in Fig. 1.6a, in proximity of a conventional s-wave SC (e.g., Al, NbTiN) in an external magnetic field [39, 40]. The Hamiltonian of Oreg-Lutchyn model reads

$$\mathcal{H} = \int dx \Big[\Psi^\dagger(x) \Big(-\frac{\hbar^2}{2m}\frac{\partial^2}{\partial x^2} - \mu - i\alpha\sigma_y \frac{\partial}{\partial x} - h\sigma_z \Big) \Psi(x) + (\Delta \psi_\uparrow(x)\psi_\downarrow(x) + H.c.) \Big]. \tag{1.15}$$

The first term of Eq. (1.15) describes the semiconducting nanowire along the x-direction with Rashba spin-orbit coupling at y-direction, while the magnetic field $\vec{B}$ should be perpendicular to the direction of spin-orbit coupling, as it is applied along the z-direction. Here, $\Psi^\dagger(x) = (\psi_\uparrow^\dagger(x), \psi_\downarrow^\dagger(x))$ with $\psi_{\uparrow(\downarrow)}^\dagger(x)$ creates an electron

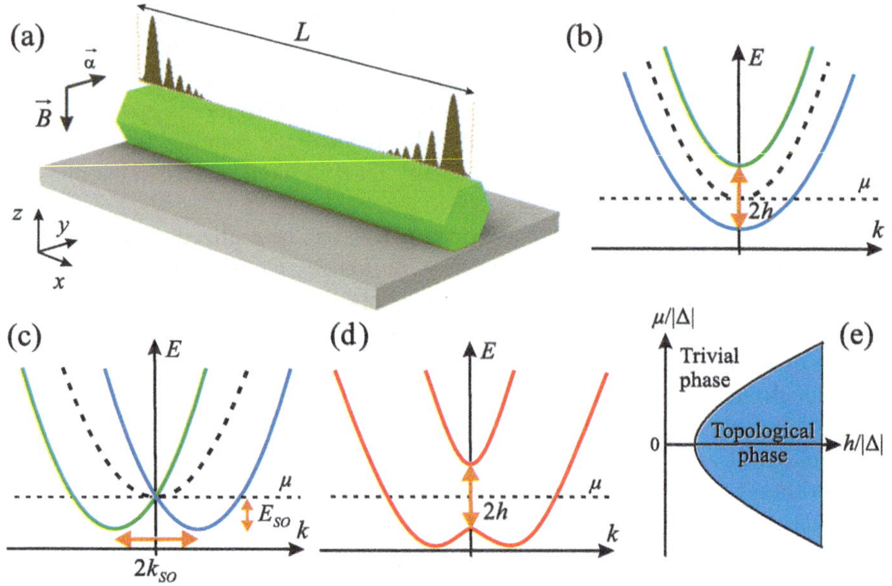

Fig. 1.6 **a** A topological superconducting nanowire (TSNW) as a result of a proximity of a semiconducting nanowire (e.g., InAs, InSb; represented in green) with strong spin-orbit coupling of strength α with an s-wave superconductor (e.g., Al, NbTiN; represented in gray) in the presence of a magnetic field ($\vec{B}$). The Majorana modes appear at the opposite ends of the TSNW (represented in orange) that overlap for a finite L length nanowire. The overlapping energy is $\varepsilon_M \propto k_F e^{-L/\xi} \cos(k_F L)$ where ξ and k_F stand for the superconducting coherence length (Majorana localization length) and the Fermi momentum, respectively. Energy dispersion of electrons in the nanowire with no proximity effect: **b** at a finite magnetic field without spin-orbit coupling where h is the Zeeman energy, **c** in the absence of a magnetic field with spin-orbit coupling (E_{SO} is the energy shift at k_{SO} induced by the spin-orbit coupling) and **d** in the presence of magnetic field and spin-orbit coupling. **e** The phase diagram as a result of proximity effect. The topological phase condition is $h > \sqrt{|\Delta|^2 + \mu^2}$ where $|\Delta|$ is the proximity induced pairing potential and μ is the chemical potential in the nanowire. Figures **a–e** adapted with permission from [24, 38]. Copyright 2017, Springer Nature. Copyright 2022 by the author. Licensed under CC BY 4.0

with spin-up (down) of effective mass m at position x. Therefore, μ represents the tunable chemical potential in the nanowire and $\alpha > 0$ denotes the spin-orbit coupling strength, which favors to align the electrons spin in y-direction [29]. The spin-orbit coupling induces an energy shift denoted by $E_{SO} = m\alpha^2/2\hbar^2$. The notations σ_y and σ_z stand for the y and z components of Pauli matrices. The quantity $h = g_W \mu_B B/2$ represents the Zeeman energy induced by the applied external magnetic field $\vec{B}$ with g_W and μ_B being the Landé factor for the nanowire and the Bohr magneton, respectively. The second term of Eq. (1.15) models the tunneling of Cooper pairs from the SC into the nanowire due to the proximity effect arising from the s-wave SC. The strength of proximity is expressed by the $\Delta = |\Delta| e^{i\varphi}$ induced superconducting pairing potential in the nanowire where φ denotes the superconducting phase. Due to the fact that the magnetic field acts unfavorably to the induced superconductivity, thus, in order to reach a strong spin-polarization in the nanowire, the Zeeman energy should be large enough. This can be realized if g_W in nanowire is very large. Appropriate candidates are the InSb or InAs semiconducting nanowires providing a large Landé-factor and a strong spin-orbit coupling [38].

To understand the physics of Hamiltonian (1.15), we first consider the case of induced superconducting pairing absence, i.e., $|\Delta| \ll \max\{h, E_{SO}\}$ [32, 38]. In the absence of a magnetic field and spin-orbit coupling, the energy dispersion of electrons is parabolic, $E(k) = \hbar^2 k^2/2m - \mu$, and is spin degenerate [see Fig. 1.6b–c, dashed line]. In the presence of magnetic field and absence of spin-orbit coupling ($h \gg E_{SO}$), the energy dispersion of Eq. (1.15) reads

$$E_{\pm}(k) = \frac{\hbar^2 k^2}{2m} - \mu \pm h, \tag{1.16}$$

which describes the motion of free electrons in a magnetic field where the two parabolas correspond to the energy of spin-$\uparrow$ and spin-$\downarrow$ electrons [see Fig. 1.6b, solid lines]. The spins are aligned with the direction of magnetic field creating thus a spin-polarization and leading to a gap of $2h$. We observe that electrons at a given energy $E(k)$ of opposite momenta $\pm k$ belong to the same spin direction.

In the absence of magnetic field and presence of a spin-orbit coupling ($h \ll E_{SO}$), the energy dispersion becomes

$$E_{\pm}(k) = \frac{\hbar^2 k^2}{2m} - \mu \pm \alpha k, \tag{1.17}$$

resulting in the two parabolic bands moved by $\pm k_{\rm SO}$ (where $k_{\rm SO} = m\alpha/\hbar^2$) from $k = 0$ on k-axis [38]. The minimum of parabolas takes the value $E_m = -(\mu + E_{\rm SO})$. We see that electrons at a given energy $E(k)$ of opposite momenta $\pm k$ belong to the opposite spin directions [see Fig. 1.6c, solid lines].

In the presence of magnetic field and spin-orbit coupling, the band energies are given by

$$E_{\pm}(k) = \frac{\hbar^2 k^2}{2m} - \mu \pm \sqrt{h^2 + (\alpha k)^2}. \tag{1.18}$$

We observe that a gap of $2h$ is opened at momentum $k = 0$. In addition, electrons with given energy $E(k)$ of opposite momenta $\pm k$ have different spin directions [see Fig. 1.6d, solid lines].

In the case when the nanowire is placed in the proximity of an s-wave SC, the last one will induce a pairing of strength Δ in the wire between electrons with opposite spin and opposite momentum [41]. The energy dispersion for quasiparticles in nanowire reads

$$E_{\pm}(k) = \sqrt{\left(\frac{\hbar^2 k^2}{2m} - \mu\right)^2 + h^2 + \alpha^2 k^2 + |\Delta|^2 \pm 2\sqrt{\left(\frac{\hbar^2 k^2}{2m} - \mu\right)^2 \cdot (h^2 + \alpha^2 k^2) + |\Delta|^2 h^2}}. \tag{1.19}$$

At momentum $k = 0$, the energy dispersion for quasiparticles in nanowire is $E_{\pm}(0) = |h \pm \sqrt{|\Delta|^2 + \mu^2}|$ which results in a gap $\Delta h = 2E_{-}(0) = 2|h - \sqrt{|\Delta|^2 + \mu^2}|$ closing at $h = \sqrt{|\Delta|^2 + \mu^2}$. Some special cases have to be distinguished. If $h < \sqrt{|\Delta|^2 + \mu^2}$, the wire is not spinless and it is in its trivial phase, producing an ordinary SC. If $h > \sqrt{|\Delta|^2 + \mu^2}$, the nanowire appears spinless and achieves its topological superconducting phase. Consequently, by tuning μ in nanowire within the gap Δh, the system can be mapped into the Kitaev model where the fermions with opposite momentum and opposite spin in the lower energy band can form p-wave pairs, driving thus the nanowire into the topological superconducting phase [36, 91]. In the last case, the semiconductor nanowire is transformed into a topological superconducting nanowire (TSNW) and at its edges MBSs emerge, a setup usually called in literature Majorana nanowire [38]. The transition between trivial and topological superconducting phases is illustrated in Fig. 1.6e.

1.3 Graphene

The carbon atom forms the basis of organic molecules and is therefore a fundamental element of life [92, 93]. They are a building block of all allotropic carbon-based substances for which the atoms can bond via different configurations providing thus a vast variety of physical properties for carbon allotropes such as fullerenes, carbon nanotubes, graphite or graphene [93, 94]. Graphene is a two-dimensional crystal for which the sp^2 hybridized carbon atoms condense in a hexagonal lattice structure, as it is shown in Fig. 1.7a. The graphene as monolayer graphite was first theoretically predicted more than a half a century [95] and isolated experimentally for almost two decades [96, 97]. The electronic band structure of monolayer graphite has been studied a long time ago within tight-binding model [95, 98, 99]. Graphene as the thinnest stretchable crystal discovered so far owns high electron mobility, good thermal conductivity, unique mechanical and optical properties [94]. The low-energy excitations in graphene are described by a two-dimensional Dirac-like equation, reason why they are often called massless Dirac fermions, which, unlike to the speed of light they move at a speed about 300 times smaller than it. Over the past years, various interesting phenomena such as half-integer quantum Hall effect [97, 100], Klein tunneling [101] and Aharonov-Bohm effect [102] have been experimentally shown in graphene. The unique physical properties of graphene make it an appropriate platform for many applications such as fabrication of electronic sensors [103, 104], optoelectronic [105] or energy storage devices [106], armors [107], membranes [108], etc. A plethora of experimental methods such as exfoliation, chemical vapor deposition, repeated peeling, epitaxial growth and sonication, etc. have been developed in order to produce graphene [94].

1.3.1 Band Structure of Graphene

1.3.1.1 The Lattice Structure and Brillouin Zone of Graphene

The monolayer graphene has a crystal structure based on a hexagonal lattice, as illustrated in Fig. 1.7a, which can be viewed as a triangular Bravais lattice with two carbon atoms per unit cell [93, 95, 109–111]. The unit cell vectors are considered as

$$\begin{aligned}\vec{a}_1 &= \frac{a_0}{2}(3, \sqrt{3}),\\ \vec{a}_2 &= \frac{a_0}{2}(3, -\sqrt{3}),\end{aligned} \tag{1.20}$$

where $a_0 \approx 1.42\text{Å}$ is the nearest-neighbor distance. The atoms of elementary cell belong to sublattices A and B, respectively. The atoms from sublattice A (B) are enclosed by three atoms from sublattice B (A). The nearest-neighbor vectors, in real space, are expressed as

$$\begin{aligned}\vec{\delta}_{11} &= \frac{a_0}{2}(1, \sqrt{3}),\\ \vec{\delta}_{12} &= \frac{a_0}{2}(1, -\sqrt{3}),\\ \vec{\delta}_{13} &= -a_0(1, 0).\end{aligned} \tag{1.21}$$

The positions of second-nearest neighbors in real space are given by the vectors $\vec{\delta}_{21} = \vec{a}_1$, $\vec{\delta}_{22} = -\vec{a}_1$, $\vec{\delta}_{23} = \vec{a}_2$, $\vec{\delta}_{24} = -\vec{a}_2$, $\vec{\delta}_{25} = \vec{a}_2 - \vec{a}_1$ and $\vec{\delta}_{26} = \vec{a}_1 - \vec{a}_2$ [not shown in Fig. 1.7a]. The vectors of reciprocal lattice are determined by using the condition $\vec{a}_i \cdot \vec{b}_j = 2\pi\delta_{ij}$. Thus, for vectors $\vec{a}_1$ and $\vec{a}_2$, one obtains

$$\begin{aligned}\vec{b}_1 &= \frac{2\pi}{3a_0}(1, \sqrt{3}),\\ \vec{b}_2 &= \frac{2\pi}{3a_0}(1, -\sqrt{3}),\end{aligned} \tag{1.22}$$

which is a triangular lattice. The reciprocal lattice of monolayer graphene is shown in Fig. 1.7b for the first Brillouin zone. The high symmetry crystallographic points, K and K', are located at the corners of the first Brillouin zone, which, as we will see

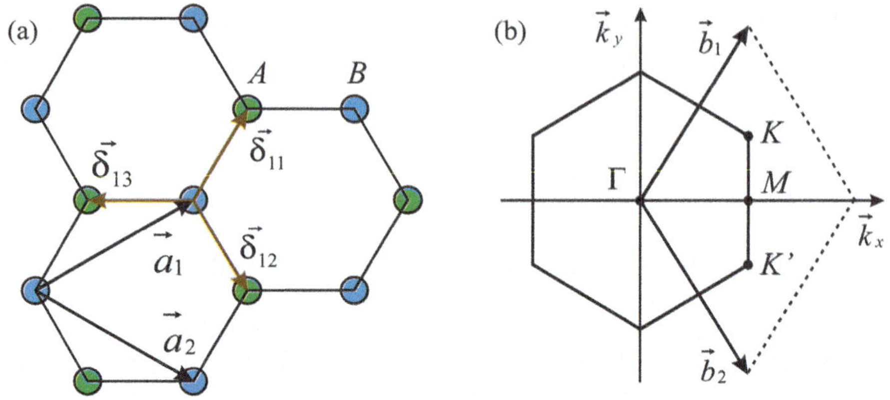

Fig. 1.7 **a** The graphene honeycomb lattice structure with basis vectors $\vec{a}_1$ and $\vec{a}_2$. The nearest-neighbor vectors $\vec{\delta}_{11}$, $\vec{\delta}_{12}$ and $\vec{\delta}_{13}$ connect the two sublattices, A and B, respectively. Adapted with permission from [93]. Copyright 2009, American Physical Society. **b** The reciprocal lattice of graphene, where the hexagon is the corresponding first Brillouin zone. The point Γ is the center of Brillouin zone and M, K and K' are inequivalent crystallographic points. Adapted with permission from [93]. Copyright 2009, American Physical Society

below, are of particular importance regarding the physical properties of graphene. The location of points K and K' in the momentum space are

$$\vec{K} = \frac{2\pi}{3a_0}\left(1, \frac{\sqrt{3}}{3}\right),$$
$$\vec{K}' = \frac{2\pi}{3a_0}\left(1, -\frac{\sqrt{3}}{3}\right). \tag{1.23}$$

The Tight-Binding Model

The unusual physical properties of graphene are attributed to its unique band structure. The carbon atoms in graphene have four valence electrons of which three are tight bonded to neighboring atoms in the graphene layer, forming thus covalent bonds. They are called σ-electrons and do not contribute to the conduction phenomena in graphene, but rather they determine its mechanical properties. The fourth electron, called π-electron, is less bonded to the carbon atom and thus it can freely move between the positive lattice ions. The unique conduction properties of graphene are determined by the π-electron. Therefore, as a first approximation, it is sufficient to consider only the delocalized electron cloud created by π electrons when determining the graphene band structure [111, 112].

Within a tight-binding approximation, considering electrons in graphene that can hop to the nearest-neighbor atoms from different sublattices (intersublattice $A-B$) with hopping energy t and to the next-nearest-neighbor atoms from same sublattices (intrasublattices $A - A$ and $B - B$) with hopping t', the energy dispersion reads [92, 93, 95, 110, 113]

$$E_{\pm}(k_x, k_y) = \pm t\sqrt{3 + \mathcal{F}(k_x, k_y)} - t'\mathcal{F}(k_x, k_y), \tag{1.24}$$

with notation

$$\mathcal{F}(k_x, k_y) = 2\cos(\sqrt{3}k_y a_0) + 4\cos\left(\frac{\sqrt{3}}{2}k_y a_0\right)\cos\left(\frac{3}{2}k_x a_0\right), \tag{1.25}$$

where the sign $+$ or $-$ refers to the conduction or valance band. The conduction and valence bands touch each other at the Dirac points. Note that for graphene the Dirac points coincide with the crystallographic K and K' points [92]. We observe that the energy dispersion becomes symmetric (electron-hole symmetry) near zero-energy for $t' = 0$. In the case of a finite $t' \neq 0$, the conduction and valance bands become asymmetric showing electron-hole asymmetry. The Dirac points have an essential role in the determination of electronic properties of graphene because around these points low-energy excitations can be achieved. For $t' = 0$, near the Dirac points, $\vec{k} = \vec{K} + \vec{q}$, where the displacement $|\vec{q}\,| \ll |\vec{K}\,|$, the energy dispersion (1.24) reduces to

$$E_{\pm}(q) \approx \pm \hbar v_F |\vec{q}\,|, \tag{1.26}$$

with $v_F = 3t\,a_0/2\hbar$ being the graphene Fermi velocity, $v_F \approx 10^6$ m/s [93]. We observe that the energy dispersion of graphene at the Dirac points for electrons/holes in the conduction/valence band is linear in momentum [as seen in Fig. 1.8a, b] presenting no energy gap [113]. Therefore, the graphene behaves as a semimetal with zero-energy gap, making it attractive for applications in physics. The linear energy band structure appearing at the Dirac points is called Dirac cone. Due to the linear energy dispersion, the charge carriers in graphene present zero effective mass, the reason why they are called massless Dirac fermions. For an intrinsic (undoped) graphene, the Fermi level is located at the Dirac point where the electron and hole bands touch each other in the momentum space. By tuning the Fermi level into the electron or hole bands, n or p doped graphene can be realized [114, 115]. The presence of Dirac points in graphene leads to a valley degeneracy, $g_v = 2$ [113].

The density of states $\rho(E)$, in the vicinity of Dirac points with $|E| \ll t$ and $t' = 0$, is given by [92, 93]

$$\rho(E) = g_s g_v \frac{|E|}{2\pi\hbar^2 v_F^2}, \tag{1.27}$$

where $g_s = 2$ is the spin degeneracy. Figure 1.8c illustrates the density of states near the Dirac point, while its inset shows the density of states for the full bandwidth. Note that the density of states disappears at $E = 0$ which is a clear consequence of the linear energy dispersion near the Dirac points [92, 93, 115, 116].

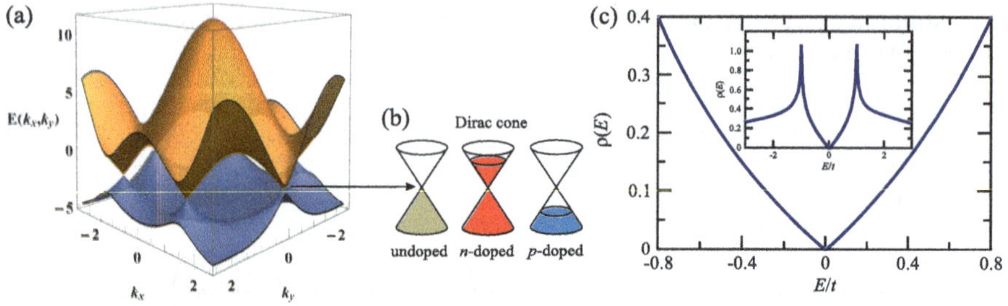

Fig. 1.8 **a** Band structure $E(k_x, k_y)$ of a monolayer graphene for parameters: $t = 2.7\,\text{eV}$ and $t' = -0.2\,t$. The valence (lower) band and conduction (upper) band touch each other at six Dirac points. Adapted with permission from [93]. Copyright 2009, American Physical Society. **b** The Dirac cones for undoped, n- and p-doped graphene appearing close to the Dirac points where the energy dispersion is linear in the momentum. Adapted from [114] under CC-BY 4.0 license. Copyright 2015 by the authors. **c** Density of states $\rho(E)$ of a monolayer graphene near the Dirac point for $t' = 0$. The inset of Figure **c** illustrates the density of states $\rho(E)$ over a full energy domain. Adapted with permission from [93, 113]. Copyrights 2009, 2011, American Physical Society

1.3.2 *Graphene Nanoribbon Geometries*

The graphene nanoribbons spectrum is strongly affected by the existence of graphene edges. There are distinguished two types of graphene nanoribbon geometries, *armchair* and *zigzag* edges, which can be realized by cutting the graphene monolayer in a semi-infinite sheet arrangement. The geometry of armchair and zigzag edges are shown in Fig. 1.9a. A graphene nanoribbon with armchair edges can be achieved by

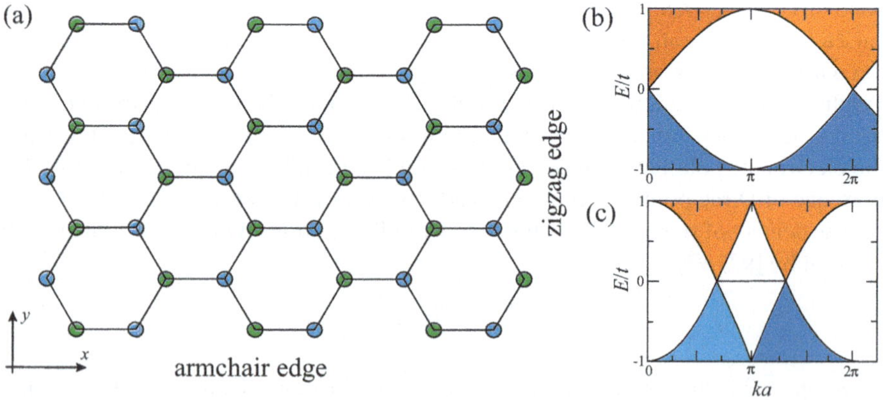

Fig. 1.9 **a** Hexagonal lattice structure showing both armchair and zigzag edges. Adapted with permission from [93]. Copyright 2009, American Physical Society. The band structure of a semi-infinite graphene nanoribbon with: **b** armchair and **c** zigzag edges. Adapted with permission from [93]. Copyright 2009, American Physical Society

considering the hexagonal lattice to be infinite in the x-direction. We can produce a graphene nanoribbon with zigzag edges by enlarging the honeycomb lattice to the infinite along the y-axis [93]. We illustrate in Fig. 1.9b, c the band structure of a semi-infinite graphene nanoribbon with armchair and zigzag edges, respectively. We observe that the energy dispersion of graphene nanoribbon with armchair edges is independent of the valley, whereas it is for zigzag edges. Note that a graphene nanoribbon with zigzag edges contains localized edge states, while graphene with armchair edges does not. The existence of localized edge states in graphene may causes its unique electric and magnetic properties [109, 117].

1.4 Optical Properties of Two-Level Quantum Systems Realized in Quantum Dots

In this section, we discuss the main phenomena determining the optical properties of QDs such as the optical intersubband transition and electric polarization of a medium for a two-level quantum system.

1.4.1 Optical Intersubband Transition

We explain the optical intersubband transition phenomenon in the case of a QD modeled by a quadratic-like potential for charge carriers as illustrated in Fig. 1.10.

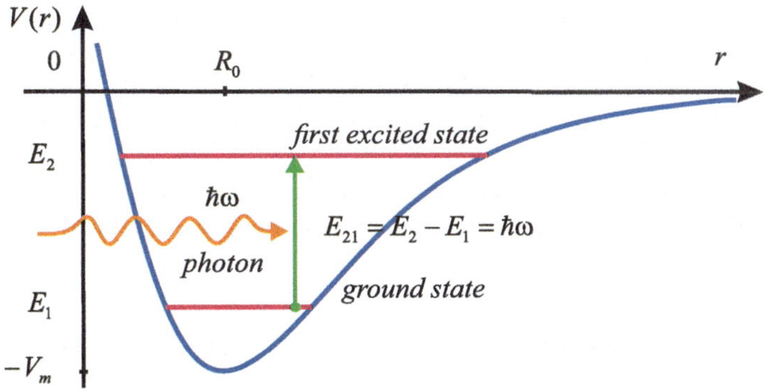

Fig. 1.10 Schematic representation of the energy levels in a QD where the confinement potential $V(r)$ for electrons in real space is arbitrary quadratic in r. Here, V_m and R_0 are the depth of potential and characteristic size of QD. The energy levels corresponding to the ground and the first excited state are denoted by E_1 and E_2, respectively. The E_{21} stands for the energy difference between the two energy levels and $\hbar\omega$ represents the energy of the incident photon. Adapted with permission from [118]. Copyright 2021, Elsevier Science & Technology Journals

The spatial confinement introduced by the potential leads to the formation of discrete energy levels. Here, E_1 and E_2 refer to the energy levels corresponding to the ground and first excited states in a QD which represent the subbands. In the case when the energy of the incident photon $\hbar\omega$ equals the energy difference between the two levels, i.e., $\hbar\omega = E_{21} = E_2 - E_1$, then an optical intersubband transition will take place. This phenomenon manifests as a photon absorption process in which an electron from ground state is excited to the higher one by irradiation of the system with an electromagnetic field, realizing thus a transition between subbands.

The absorption of electromagnetic energy refers to the process when the bounded electrons in an irradiated material transform the energy of photons into internal energy. The energy transferred to a material at a given frequency is expressed by the optical absorption coefficient. When the incident photon energy matches the energy difference between subbands, a resonance in the optical absorption spectrum appears [119, 120]. Another relevant optical parameter of materials is the refractive index, which is the ratio between the propagation speed of light measured in vacuum and in the material. Due to the dispersion phenomenon, the refractive index of a medium depends on wavelength of incident light, on the photon energy [121, 122]. Here, we are interested in the relative changes in refractive index of a material varying the QD characteristic size, the potential depth and the incident optical intensity.

1.4.2 Electric Susceptibilities

The optical properties of QDs, characterized by the optical absorption coefficient and refractive index of medium, can be analyzed within the framework of a compact density-matrix approach combined with perturbation technique [119, 122–124]. The applied electromagnetic field to the QD system, corresponding to the incident excitation photons, is polarized along the z-axis. Thus, the time-dependent electric component of the electromagnetic field is expressed as $E(t) = \tilde{E}e^{i\omega t} + \tilde{E}^*e^{-i\omega t}$ where $\tilde{E}$ is the amplitude of electric field, while ω stands for its angular frequency [119, 125].

The electric polarization $\mathcal{P}(t)$, meaning the density of induced electric dipole moments in a material upon applying an electric field, is calculated as [119, 120, 126, 127]

$$\mathcal{P}(t) = \varepsilon_0 \chi(\omega)\tilde{E}e^{i\omega t} + \varepsilon_0 \chi(-\omega)\tilde{E}^*e^{-i\omega t} = \frac{1}{\mathcal{V}}\mathrm{Tr}(\rho\mathcal{M}), \qquad (1.28)$$

where ρ and $\mathcal{M}$ stand for the one-electron density operator and the electric dipole operator along the z-axis. Here it is implicitly assumed that the electric polarization vanishes in the absence of the external electromagnetic field. The applied electromagnetic field induces charge separation, which leads to the formation of electric dipole moments (i.e., polarization of the material). Here, $\chi(\omega)$ and $\mathcal{V}$ are the electric susceptibility of the medium and the volume of the system with ε_0 being the permittivity of vacuum. The electric susceptibility expresses the response of a medium to an applied electric field. For small values of the electric field strength $E(t)$, the electric polarization $\mathcal{P}(t)$ given by Eq. (1.28) can be expanded as a power series of $E(t)$ up to the third-order [120, 122, 124, 128]. One finds

$$\begin{aligned}\mathcal{P}(t) \approx \varepsilon_0\big[&\chi^{(1)}(\omega)\tilde{E}e^{i\omega t} + \chi_0^{(2)}|\tilde{E}|^2 + \chi^{(2)}(2\omega)\tilde{E}^2e^{i2\omega t}\\ &+ \chi^{(3)}(\omega)|\tilde{E}|^2\tilde{E}e^{i\omega t} + \chi^{(3)}(3\omega)\tilde{E}^3e^{i3\omega t} + ... + c.c.\big],\end{aligned} \tag{1.29}$$

where $\chi^{(1)}(\omega)$, $\chi_0^{(2)}$, $\chi^{(2)}(2\omega)$, $\chi^{(3)}(\omega)$, and $\chi^{(3)}(3\omega)$ are the first-order (linear), optical rectification, second harmonic generation, third-order nonlinear and third harmonic generation susceptibilities. Note that even orders of $\chi(\omega)$, i.e., the optical rectification and second harmonic generation terms of the susceptibility vanish due to the inversion symmetry of the system [120, 122, 124]. By neglecting the higher harmonic terms of $\chi(\omega)$ in Eq. (1.29), the total susceptibility is expressed as the sum of linear and third-order nonlinear susceptibilities. Based on the compact density-matrix approach and using perturbation theory, the linear $\chi^{(1)}(\omega)$ and third-order nonlinear $\chi^{(3)}(\omega)$ terms of the total electric susceptibility are computed as [119, 122–125, 127]

$$\varepsilon_0\chi^{(1)}(\omega) = \frac{\sigma_\nu|\mathcal{M}_{21}|^2}{E_{21} - \hbar\omega - i\hbar\Gamma_{12}}, \tag{1.30}$$

and

$$\begin{aligned}\varepsilon_0\chi^{(3)}(\omega) = -&\frac{\sigma_\nu|\mathcal{M}_{21}|^2}{E_{21} - \hbar\omega - i\hbar\Gamma_{12}}\left[\frac{4|\mathcal{M}_{21}|^2}{(E_{21} - \hbar\omega)^2 + (\hbar\Gamma_{12})^2}\right.\\ &\left.- \frac{|\mathcal{M}_{22} - \mathcal{M}_{11}|^2}{(E_{21} - i\hbar\Gamma_{12})(E_{21} - \hbar\omega - i\hbar\Gamma_{12})}\right],\end{aligned} \tag{1.31}$$

respectively. The term $\mathcal{M}_{ij} = \langle\psi_{n_i\ell_i m_i}|ez|\psi_{n_j\ell_j m_j}\rangle$ represents the matrix elements of the electric dipole moment, $\mathcal{M}$, where j denotes the energy level for quantum state $|\psi_{n_j\ell_j m_j}\rangle$ with principal n_j, orbital ℓ_j, and magnetic m_j quantum numbers. The matrix element $\mathcal{M}_{ij}$ is proportional to the one-electron optical transition probability amplitude between states $|\psi_{n_j\ell_j m_j}\rangle$ and $|\psi_{n_i\ell_i m_i}\rangle$, respectively, induced by the absorption of a single photon. In addition, $E_{ij} = E_i - E_j$ refers to the energy difference between energy levels corresponding to quantum states $|\psi_{n_i\ell_i m_i}\rangle$ and $|\psi_{n_j\ell_j m_j}\rangle$, respectively. The non-diagonal matrix element $\Gamma_{ij} = 1/T_{ij}$ (with $i \neq j$) represents the relaxation rate of transition between the states $|\psi_{n_i\ell_i m_i}\rangle$ and $|\psi_{n_j\ell_j m_j}\rangle$ by defining a relaxation time T_{ij} [119, 129–133]. For specific systems, the values of the relaxation times T_{ij} are taken from the literature. Recall that the notation (2)1 in Eqs. (1.30)–(1.31) refers to the (final) initial state and $\hbar\omega$ symbolizes the energy of the incident photon. The carrier density is denoted by σ_ν.

References

1. L.P. Kouwenhoven, C.M. Marcus, P.L. McEuen, S. Tarucha, R.M. Westervelt, N.S. Wingreen, *Electron Transport in Quantum Dots* (Springer Netherlands, Dordrecht, 1997), pp. 105–214
2. L.P. Kouwenhoven, P.L. McEuen, *Single Electron Transport Through a Quantum Dot.* (Springer, New York, 1999), pp. 471–535
3. S. Datta, *Quantum Transport: Atom to Transistor* (Cambridge University Press, 2005)
4. N. Singh, Quantum transport in mesoscopic systems. Resonance **15**, 988 (2010)
5. Y.D. Baines, *Kondo physics in side coupled quantum dots* (Université Joseph-Fourier - Grenoble I, Theses, Dec.2010)
6. C.W.J. Beenakker, Theory of Coulomb-blockade oscillations in the conductance of a quantum dot. Phys. Rev. B **44**, 1646–1656 (1991)
7. D. Ryndyk, *Theory of Quantum Transport at Nanoscale: An Introduction* (Springer International Publishing, Cham, Switzerland, 2016)
8. C. Führer, *Magneto-transport investigations on multi-electron quantum dots: Coulomb blockade, Kondo effect, and Fano regime*. Ph.D. dissertation, Hannover: Universität (2002)
9. S. De Franceschi, W. van der Wiel, *Kondo Effect in Quantum Dots* (CRC Press, (Taylor & Francis), 2011), pp. 646–664
10. R. Van Roermund, *Theoretical Study of non-equilibrium transport in Kondo quantum dots* (Université de Grenoble, Theses, 2010)
11. M.A. Reed, J.N. Randall, R.J. Aggarwal, R.J. Matyi, T.M. Moore, A.E. Wetsel, Observation of discrete electronic states in a zero-dimensional semiconductor nanostructure. Phys. Rev. Lett. **60**, 535–537 (1988)
12. D. Goldhaber-Gordon, H. Shtrikman, D. Mahalu, D. Abusch-Magder, U. Meirav, M.A. Kastner, Kondo effect in a single-electron transistor. Nature **391**, 156–159 (1998)
13. S.M. Cronenwett, T.H. Oosterkamp, L.P. Kouwenhoven, A tunable Kondo effect in quantum dots. Science **281**(5376), 540–544 (1998)
14. J. Schmid, J. Weis, K. Eberl, K.v. Klitzing, A quantum dot in the limit of strong coupling to reservoirs. Phys. B **256-258**, 182–185 (1998)
15. W.G. van der Wiel, S.D. Franceschi, T. Fujisawa, J.M. Elzerman, S. Tarucha, L.P. Kouwenhoven, The Kondo effect in the unitary limit. Science **289**(5487), 2105–2108 (2000)
16. L. Kouwenhoven, L. Glazman, Revival of the Kondo effect. Phys. World **14**, 33 (2001)
17. J. Sólyom, *Fundamentals of the Physics of Solids: Electronic Properties* (Springer, Berlin Heidelberg, 2009)
18. W. de Haas, J. de Boer, and G. van den Berg, The electrical resistance of gold, copper and lead at low temperature. Physica **1**(7,) 1115–1124 (1934)
19. J. Kondo, Resistance minimum in dilute magnetic alloys. Prog. Theor. Phys. **32**, 37–49, 07 (1964)
20. P. Nozières, A "fermi-liquid" description of the Kondo problem at low temperatures. J. Low Temp. Phys. **17**, 31 (1974)
21. K.G. Wilson, The renormalization group: Critical phenomena and the Kondo problem. Rev. Mod. Phys. **47**, 773–840 (1975)
22. P.W. Anderson, Localized magnetic states in metals. Phys. Rev. **124**, 41–53 (1961)
23. M.A. Sierra, R. López, D. Sánchez, Fate of the spin-$\frac{1}{2}$ Kondo effect in the presence of temperature gradients. Phys. Rev. B **96**, 085416 (2017)
24. R. Aguado, Majorana quasiparticles in condensed matter. Riv. Nuovo Cim. **40**(11), 523–593 (2017)
25. R. Hützen, *Transport through interacting quantum dots with Majorana fermions or phonons.* PhD dissertation, Heinrich-Heine-University Düsseldorf (2013)
26. E. Majorana, Teoria simmetrica dell'elettrone e del positrone. Nuovo Cim. **14**, 171 (1937)
27. S.R. Elliott, M. Franz, Colloquium: Majorana fermions in nuclear, particle, and solid-state physics. Rev. Mod. Phys. **87**, 137–163 (F2015)
28. C.D. Anderson, The positive electron. Phys. Rev. **43**, 491–494 (1933)

29. D. Sticlet, *Edge states in Chern Insulators and Majorana fermions in topological superconductors* (Université Paris Sud - Paris XI, Theses, 2012)
30. F. Wilczek, Majorana returns. Nat. Phys. **5**, 614 (2009)
31. C. Beenakker, Search for Majorana fermions in superconductors. Annu. Rev. Condens. Matter. Phys. **4**, 113–136 (2013)
32. F.V. Oppen, Y. Peng, F. Pientka, Topological superconducting phases in one dimension, in *Topological Aspects of Condensed Matter Physics: Lecture Notes of the Les Houches Summer School: Volume 103, August 2014* (Oxford University Press, 01 2017)
33. A.Y. Kitaev, Unpaired Majorana fermions in quantum wires. Phys. Usp. **44**, 131–136 (2001)
34. A. Kitaev, Anyons in an exactly solved model and beyond. Ann. Phys. **321**, 2–111 (2006)
35. C. Nayak, S. H. Simon, A. Stern, M. Freedman, S. Das Sarma, Non-Abelian anyons and topological quantum computation. Rev. Mod. Phys. **80**, 083–1159 (2008)
36. J. Alicea, New directions in the pursuit of Majorana fermions in solid state systems. Rep. Prog. Phys. **75**, 07650 (2012)
37. M. Leijnse, K. Flensberg, Introduction to topological superconductivity and Majorana fermions. Semicond. Sci. Technol. **27**, 124003 (2012)
38. P. Marra, Majorana nanowires for topological quantum computation. J. Appl. Phys. **132**, 231101,12 (2022)
39. Y. Oreg, G. Refael, F. von Oppen, Helical liquids and Majorana bound states in quantum wires. Phys. Rev. Lett. **105**, 177002 (2010)
40. R. M. Lutchyn, J. D. Sau, S. Das Sarma, Majorana fermions and a topological phase transition in semiconductor-superconductor heterostructures. Phys. Rev. Lett. **105**, 077001 (2010)
41. V. Mourik, K. Zuo, S.M. Frolov, S.R. Plissard, E.P.A.M. Bakkers, L.P. Kouwenhoven, Signatures of Majorana fermions in hybrid superconductor-semiconductor nanowire devices. Science **336**(6084), 1003 (2012)
42. A. Das, Y. Ronen, Y. Most, Y. Oreg, M. Heiblum, H. Shtrikman, Zero-bias peaks and splitting in an Al–InAs nanowire topological superconductor as a signature of Majorana fermions. Nat. Phys. **8**, 887–895 (2012)
43. M.T. Deng, C.L. Yu, G.Y. Huang, M. Larsson, P. Caroff, H.Q. Xu, Anomalous zero-bias conductance peak in a Nb–InSb nanowire–Nb hybrid device. Nano Lett. **12**, 6414–6419 (2012)
44. A.D.K. Finck, D.J. Van Harlingen, P.K. Mohseni, K. Jung, X. Li, Anomalous modulation of a zero-bias peak in a hybrid nanowire-superconductor device. Phys. Rev. Lett. **110**, 126406 (2013)
45. H.O.H. Churchill, V. Fatemi, K. Grove-Rasmussen, M.T. Deng, P. Caroff, H.Q. Xu, C.M. Marcus, Superconductor-nanowire devices from tunneling to the multichannel regime: Zero-bias oscillations and magnetoconductance crossover. Phys. Rev. B **87**, 241401 (2013)
46. N. Read, D. Green, Paired states of fermions in two dimensions with breaking of parity and time-reversal symmetries and the fractional quantum Hall effect. Phys. Rev. B **61**, 10267–10297 (2000)
47. M. Stone, R. Roy, Edge modes, edge currents, and gauge invariance in $p_x + ip_y$ superfluids and superconductors. Phys. Rev. B **69**, 184511 (2004)
48. L. Fu, C.L. Kane, Superconducting proximity effect and Majorana fermions at the surface of a topological insulator. Phys. Rev. Lett. **100**, 096407 (2008)
49. S. Nadj-Perge, I.K. Drozdov, J. Li, H. Chen, S. Jeon, J. Seo, A.H. MacDonald, B.A. Bernevig, A. Yazdani, Observation of Majorana fermions in ferromagnetic atomic chains on a superconductor. Science **346**, 602–607 (2014)
50. M. Ruby, F. Pientka, Y. Peng, F. von Oppen, B.W. Heinrich, K.J. Franke, End states and subgap structure in proximity-coupled chains of magnetic adatoms. Phys. Rev. Lett. **115**, 197204 (2015)
51. R. Pawlak, M. Kisiel, J. Klinovaja, T. Meier, S. Kawai, T. Glatzel, D. Loss, E. Meyer, Probing atomic structure and Majorana wavefunctions in mono-atomic Fe chains on superconducting Pb surface. npj Quantum Inf. **2**, 16035 (2016)

52. B.E. Feldman, M.T. Randeria, J. Li, S. Jeon, Y. Xie, Z. Wang, I.K. Drozdov, B. Andrei Bernevig, A. Yazdani, High-resolution studies of the Majorana atomic chain platform. Nat. Phys. **13**, 286 (2017)
53. S. Jeon, Y. Xie, J. Li, Z. Wang, B.A. Bernevig, A. Yazdani, Distinguishing a Majorana zero mode using spin-resolved measurements. Science **358**(6364), 772–776 (2017)
54. H. Kim, A. Palacio-Morales, T. Posske, L. Rózsa, K. Palotás, L. Szunyogh, M. Thorwart, R. Wiesendanger, Toward tailoring Majorana bound states in artificially constructed magnetic atom chains on elemental superconductors. Sci. Adv. **4** (2018)
55. L. Schneider, P. Beck, T. Posske, D. Crawford, E. Mascot, S. Rachel, R. Wiesendanger, J. Wiebe, Topological Shiba bands in artificial spin chains on superconductors. Nat. Phys. **17**, 943–948 (2021)
56. L. Schneider, P. Beck, J. Neuhaus-Steinmetz, L. Rózsa, T. Posske, J. Wiebe, R. Wiesendanger, Precursors of Majorana modes and their length-dependent energy oscillations probed at both ends of atomic Shiba chains. Nat. Nanotechnol. **17**, 384–389 (2022)
57. M. Duckheim, P.W. Brouwer, Andreev reflection from noncentrosymmetric superconductors and Majorana bound-state generation in half-metallic ferromagnets. Phys. Rev. B **83**, 054513 (2011)
58. S.B. Chung, H.-J. Zhang, X.-L. Qi, S.-C. Zhang, Topological superconducting phase and Majorana fermions in half-metal/superconductor heterostructures. Phys. Rev. B **84**, 060510 (2011)
59. T.-P. Choy, J.M. Edge, A.R. Akhmerov, C.W.J. Beenakker, Majorana fermions emerging from magnetic nanoparticles on a superconductor without spin-orbit coupling. Phys. Rev. B **84**, 195442 (2011)
60. M. Kjaergaard, K. Wölms, K. Flensberg, Majorana fermions in superconducting nanowires without spin-orbit coupling. Phys. Rev. B **85**, 020503 (2012)
61. L. Jiang, T. Kitagawa, J. Alicea, A.R. Akhmerov, D. Pekker, G. Refael, J.I. Cirac, E. Demler, M.D. Lukin, P. Zoller, Majorana fermions in equilibrium and in driven cold-atom quantum wires. Phys. Rev. Lett. **106**, 220402 (2011)
62. J.-P. Xu, M.-X. Wang, Z.L. Liu, J.-F. Ge, X. Yang, C. Liu, Z.A. Xu, D. Guan, C.L. Gao, D. Qian, Y. Liu, Q.-H. Wang, F.-C. Zhang, Q.-K. Xue, J.-F. Jia, Experimental detection of a Majorana mode in the core of a magnetic vortex inside a topological insulator-superconductor $bi_2te_3/nbse_2$ heterostructure. Phys. Rev. Lett. **114**, 017001 (2015)
63. D. Wang, L. Kong, P. Fan, H. Chen, S. Zhu, W. Liu, L. Cao, Y. Sun, S. Du, J. Schneeloch, R. Zhong, G. Gu, L. Fu, H. Ding, H.-J. Gao, Evidence for Majorana bound states in an iron-based superconductor. Science **362**, 333–335 (2018)
64. G.C. Ménard, A. Mesaros, C. Brun, F. Debontridder, D. Roditchev, P. Simon, T. Cren, Isolated pairs of Majorana zero modes in a disordered superconducting lead monolayer. Nat. Commun. **10** (2019)
65. C.-K. Chiu, T. Machida, Y. Huang, T. Hanaguri, F.-C. Zhang, Scalable Majorana vortex modes in iron-based superconductors. Sci. Adv. **6** (2020)
66. Z. Cao, D.E. Liu, W.-X. He, X. Liu, K. He, H. Zhang, Numerical study of PbTe-Pb hybrid nanowires for engineering Majorana zero modes. Phys. Rev. B **105**, 085424 (2022)
67. E.J.H. Lee, X. Jiang, R. Aguado, G. Katsaros, C.M. Lieber, S. De Franceschi, Zero-bias anomaly in a nanowire quantum dot coupled to superconductors. Phys. Rev. Lett. **109**, 186802 (2012)
68. R. Žitko, J.S. Lim, R. López, R. Aguado, Shiba states and zero-bias anomalies in the hybrid normal-superconductor Anderson model. Phys. Rev. B **91**, 045441 (2015)
69. A. Vuik, B. Nijholt, A. Akhmerov, M. Wimmer, Reproducing topological properties with quasi-Majorana states. SciPost Phys. **7** (2019)
70. H. Pan, C.-X. Liu, M. Wimmer, S. Das Sarma, Quantized and unquantized zero-bias tunneling conductance peaks in majorana nanowires: Conductance below and above $2e^2/h$. Phys. Rev. B **103**, 214502 (2021)
71. G. Kells, D. Meidan, P.W. Brouwer, Near-zero-energy end states in topologically trivial spin-orbit coupled superconducting nanowires with a smooth confinement. Phys. Rev. B **86**, 100503 (2012)

72. M. Franz, Majorana's wires. Nat. Nanotechnol. **8**, 149–152 (2013)
73. E. Prada, P. San-Jose, M.W.A. de Moor, A. Geresdi, E.J.H. Lee, J. Klinovaja, D. Loss, J. Nygard, R. Aguado, L.P. Kouwenhoven, From Andreev to Majorana bound states in hybrid superconductor-semiconductor nanowires. Nat. Rev. Phys. **2**, 575–594 (2020)
74. J. Liu, A.C. Potter, K.T. Law, P.A. Lee, Zero-bias peaks in the tunneling conductance of spin-orbit-coupled superconducting wires with and without Majorana end-states. Phys. Rev. Lett. **109**, 267002 (2012)
75. D.I. Pikulin, J.P. Dahlhaus, M. Wimmer, H. Schomerus, C.W.J. Beenakker, A zero-voltage conductance peak from weak antilocalization in a Majorana nanowire. New J. Phys. **14**, 125011 (2012)
76. S. Das Sarma, H. Pan, Disorder-induced zero-bias peaks in Majorana nanowires. Phys. Rev. B **103**, 195158 (2021)
77. H. Pan, J. D. Sau, S. Das Sarma, Three-terminal nonlocal conductance in Majorana nanowires: Distinguishing topological and trivial in realistic systems with disorder and inhomogeneous potential. Phys. Rev. B **103**, 014513 (2021)
78. H. Pan, S. Das Sarma, Disorder effects on Majorana zero modes: Kitaev chain versus semiconductor nanowire. Phys. Rev. B **103**, 224505 (2021)
79. H. Pan, S. Das Sarma, Crossover between trivial zero modes in Majorana nanowires. Phys. Rev. B **104**, 054510 (2021)
80. E. Prada, P. San-Jose, R. Aguado, Transport spectroscopy of *ns* nanowire junctions with Majorana fermions. Phys. Rev. B **86**, 180503 (2012)
81. C. Moore, C. Zeng, T.D. Stanescu, S. Tewari, Quantized zero-bias conductance plateau in semiconductor-superconductor heterostructures without topological majorana zero modes. Phys. Rev. B **98**, 155314 (2018)
82. T.D. Stanescu, S. Tewari, Robust low-energy andreev bound states in semiconductor-superconductor structures: Importance of partial separation of component majorana bound states. Phys. Rev. B **100**, 155429 (2019)
83. L. Fu, C.L. Kane, Josephson current and noise at a superconductor/quantum-spin-Hall-insulator/superconductor junction. Phys. Rev. B **79**, 161408 (2009)
84. L. Jiang, D. Pekker, J. Alicea, G. Refael, Y. Oreg, F. von Oppen, Unconventional Josephson signatures of Majorana bound states. Phys. Rev. Lett. **107**, 236401 (2011)
85. J. Nilsson, A.R. Akhmerov, C.W.J. Beenakker, Splitting of a Cooper pair by a pair of Majorana bound states. Phys. Rev. Lett. **101**, 120403 (2008)
86. K.T. Law, P.A. Lee, T.K. Ng, Majorana fermion induced resonant Andreev reflection. Phys. Rev. Lett. **103**, 237001 (2009)
87. K. Flensberg, Tunneling characteristics of a chain of Majorana bound states. Phys. Rev. B **82**, 180516 (2010)
88. N. Leumer, M. Grifoni, B. Muralidharan, M. Marganska, Linear and nonlinear transport across a finite Kitaev chain: an exact analytical study. Phys. Rev. B **103**, 165432 (2021)
89. C. Duse, P. Sriram, K. Gharavi, J. Baugh, B. Muralidharan, Role of dephasing on the conductance signatures of Majorana zero modes. J. Phys.: Condens. Matter. **33**, 365301 (2021)
90. V. Lahtinen, J.K. Pachos, A short introduction to topological quantum computation. SciPost Phys. **3**, 021 (2017)
91. J. Alicea, Y. Oreg, G. Refael, F. von Oppen, M.P.A. Fisher, Non-Abelian statistics and topological quantum information processing in 1d wire networks. Nat. Phys. **7**, 412 (2011)
92. J.-N. Fuchs M. Goerbig, Introduction to the physical properties of graphene, in *Lecture notes* (2008)
93. A.H. Castro Neto, F. Guinea, N.M.R. Peres, K.S. Novoselov, A.K. Geim, The electronic properties of graphene. Rev. Mod. Phys. **81**, 109–162 (2009)
94. R. Tarcan, O. Todor-Boer, I. Petrovai, C. Leordean, S. Astilean, I. Botiz, Reduced graphene oxide today. J. Mater. Chem. C **8**, 1198–1224 (2020)
95. P.R. Wallace, The band theory of graphite. Phys. Rev. **71**, 622–634 (1947)
96. K.S. Novoselov, A.K. Geim, S.V. Morozov, D. Jiang, Y. Zhang, S.V. Dubonos, I.V. Grigorieva, A.A. Firsov, Electric field effect in atomically thin carbon films. Science **306**, 666–669 (2004)

97. K.S. Novoselov, A.K. Geim, S.V. Morozov, D. Jiang, M.I. Katsnelson, I.V. Grigorieva, S.V. Dubonos, A.A. Firsov, Two-dimensional gas of massless Dirac fermions in graphene. Nature **438**, 197–200 (2005)
98. J.W. McClure, Band structure of graphite and de Haas-van Alphen effect. Phys. Rev. **108**, 612–618 (1957)
99. J.C. Slonczewski, P.R. Weiss, Band structure of graphite. Phys. Rev. **109**, 272–279 (1958)
100. Y. Zhang, Y.-W. Tan, H.L. Stormer, P. Kim, Experimental observation of the quantum Hall effect and Berry's phase in graphene. Nature **438**, 201 (2005)
101. N. Stander, B. Huard, D. Goldhaber-Gordon, Evidence for Klein tunneling in graphene $p-n$ junctions. Phys. Rev. Lett. **102**, 026807 (2009)
102. S. Russo, J.B. Oostinga, D. Wehenkel, H.B. Heersche, S.S. Sobhani, L.M.K. Vandersypen, A.F. Morpurgo, Observation of Aharonov-Bohm conductance oscillations in a graphene ring. Phys. Rev. B **77**, 085413 (2008)
103. Q. He, S. Wu, Z. Yin, H. Zhang, Graphene-based electronic sensors. Chem. Sci. **3**, 1764–1772 (2012)
104. M. Xu, J. Qi, F. Li, X. Liao, S. Liu, Y. Zhang, Ultra-thin, transparent and flexible tactile sensors based on graphene films with excellent anti-interference. RSC Adv. **7**, 30506–30512 (2017)
105. K. Rana, J. Singh, J.-H. Ahn, A graphene-based transparent electrode for use in flexible optoelectronic devices. J. Mater. Chem. C **2**, 2646–2656 (2014)
106. A. Olabi, M.A. Abdelkareem, T. Wilberforce, E.T. Sayed, Application of graphene in energy storage device – A review. Renew. Sustain. Energy Rev. **135**, 110026 (2021)
107. J.-H. Lee, P.E. Loya, J. Lou, E.L. Thomas, Dynamic mechanical behavior of multilayer graphene via supersonic projectile penetration. Science **346**(6213), 1092–1096 (2014)
108. V. Berry, Impermeability of graphene and its applications. Carbon **62**, 1–10 (2013)
109. J.J. Palacios, J. Fernández-Rossier, L. Brey, H.A. Fertig, Electronic and magnetic structure of graphene nanoribbons. Semicond. Sci. Technol. **25**, 033003 (2010)
110. M. Katsnelson, *Graphene: Carbon in Two Dimensions* (Cambridge University Press, 2012)
111. G.Z. Magda, *Grafn nanoszerkezetek s ms ktdimenzis anyagok kialaktsa s vizsglata psztzszonds mdszerekkel*. Ph.D. dissertation, Budapest University of Technology and Economics (2017)
112. A. Maffucci, G. Miano, Electrical properties of graphene for interconnect applications. Appl. Sci. **4**(2), 305–317 (2014)
113. S. Das Sarma, S. Adam, E. H. Hwang, E. Rossi, Electronic transport in two-dimensional graphene. Rev. Mod. Phys. **83**, 407–470 (2011)
114. H. Wang, H. Zhao, G. Hu, S. Li, H. Su, J. Zhang, Graphene based surface plasmon polariton modulator controlled by ferroelectric domains in lithium niobate. Sci. Rep. **5**, 18258 (2015)
115. C.W.J. Beenakker, Colloquium: Andreev reflection and Klein tunneling in graphene. Rev. Mod. Phys. **80**, 1337–1354 (2008)
116. E. Hwang, B.Y.K. Hu, S. Das Sarma, Many-body exchange-correlation effects in graphene. Physica E **40**(5), 1653–1655 (2008)
117. K. Wakabayashi, S. Dutta, Nanoscale and edge effect on electronic properties of graphene. Solid State Commun. **152**(15), 1420–1430 (2012). Exploring Graphene, Recent Research Advances
118. L. Máthé, C. Onyenegecha, A.-A. Farcaş, L.-M. Pioraş-Ţimbolmaş, M. Solaimani, H. Hassanabadi, Linear and nonlinear optical properties in spherical quantum dots: Inversely quadratic Hellmann potential. Phys. Lett. A **397**, 127262 (2021)
119. D. Ahn, S. lien Chuang, Calculation of linear and nonlinear intersubband optical absorptions in a quantum well model with an applied electric field. IEEE J. Quantum Electron. **23**(12), 1196–2204 (1987)
120. K.J. Kuhn, G.U. Iyengar, S. Yee, Free carrier induced changes in the absorption and refractive index for intersubband optical transitions in $Al_xGa_{1-x}As/GaAs/Al_xGa_{1-x}As$ quantum wells. J. Appl. Phys. **70**(9), 5010–5017 (1991)
121. E. Hecht, *Optics* (Addison Wesley, San Francisco, 4th ed., 2002)
122. R.W. Boyd, *Nonlinear Optics, Third Edition* (Academic Press, Inc., USA, 3rd ed., 2008)

123. S. Ünlü, I. Karabulut, H. Safak, Linear and nonlinear intersubband optical absorption coefficients and refractive index changes in a quantum box with finite confining potential. Physica E **33**(2), 319–324 (2006)
124. M. Vahdani, The effect of the electronic intersubband transitions of quantum dots on the linear and nonlinear optical properties of dot-matrix system. Superlattices Microstruct. **76**, 326–338 (2014)
125. G. Liu, K. Guo, Linear and nonlinear intersubband optical absorption coefficients and refractive index changes with the ring-shaped non-spherical oscillator potential. Superlattices Microstruct. **52**(5), 997–1009 (2012)
126. R. Khordad, Optical properties of quantum wires: Rashba effect and external magnetic field. J. Lumin. **134**, 201–207 (2013)
127. R. Khordad, B. Mirhosseini, Linear and nonlinear optical properties in spherical quantum dots: Rosen-Morse potential. Opt. Spectrosc. **117**, 434–440 (2014)
128. U.e. Kalsoom, R. Yi, J. Qu, L. Liu, Nonlinear optical properties of CdSe and CdTe core-shell quantum dots and their applications. Front. Phys. (Lausanne) **9** (2021)
129. L. Lu, W. Xie, H. Hassanabadi, The effects of intense laser on nonlinear properties of shallow donor impurities in quantum dots with the Woods-Saxon potential. J. Lumin. **131**(12), 2538–2543 (2011)
130. F. Ungan, M.K. Bahar, S. Pal, M.E. Mora-Ramos, Electron-related nonlinear optical properties of cylindrical quantum dot with the Rosen–Morse axial potential. Commun. Theor. Phys. **72**, 075505 (2020)
131. H. El Ghazi, A. Jorio, I. Zorkani, Linear and nonlinear intra-conduction band optical absorption in (In, Ga)N/GaN spherical QD under hydrostatic pressure. Opt. Commun. **331**, 73–76 (2014)
132. Y. Yakar, B. Çakir, A. Özmen, Linear and nonlinear absorption coefficients of spherical two-electron quantum dot. Comput. Phys. Commun. **188**, 88–93 (2015)
133. D. Kaya, M. Ycel, O. Bayrak, The total absorption coefficient and refractive index change of a spherical quantum dot in the confinement radial potential. Physica B **648**, 414403 (2023)

Chapter 2
Probing Majorana Bound States via Quantum Dots in the Presence of Electron-Phonon Interaction

This chapter of the book proposes an experimentally feasibly nanojunction in which a quantum dot is integrated in a semiconducting-superconducting heterostructure hosting Majorana bound states (MBSs). The realistic setup we propose could enable the probing of Majorana modes where the energy exchange of the nanodevice with an environmental bath is also taken into account within a single long-wave optical phonon mode. A threading magnetic flux is also considered to manipulate the Majorana-induced phonon-assisted quantum tunneling. The electron-phonon interaction (EPI) is incorporated into the model employing a canonical transformation combined with the nonequilibrium Green's function technique. The results obtained in this chapter can serve as a promising guide in Majorana detection experiments. Furthermore, the device investigated here is highly relevant to research on the implementation of topological quantum computation.

2.1 State-of-The-Art and Motivation

The discovery of MBSs has opened a new research area in the condensed matter community. The most commonly employed scheme to probe MBSs in a topological superconducting nanowire (TSNW) requires embedding a nanowire in a mesoscopic circuit. A minimal setup is realized from Majorana zero modes (MZMs) connected to regular fermionic degrees of freedom associated with a quantum dot (QD), which is then coupled to electrodes [1–3]. The signature of MZMs is seen in the linear conductance measured through the dot [2]. The theoretical investigations on transport properties of a noninteracting QD, in the on-resonance with symmetrically connected to normal leads and attached to a TSNW, reported that the Majorana signature drastically affects the linear conductance [2]. Namely, when the QD connects to a regular fermionic zero mode, the linear conductance is 0. When the QD is connected to

© The Author(s), under exclusive license to Springer Nature Switzerland AG 2025
L. Máthé, *Physical Properties of Low-Dimensional Systems*,
NanoScience and Technology, https://doi.org/10.1007/978-3-031-91602-1_2

a nanowire which is in its topologically trivial phase, the linear conductance peak takes the value e^2/h, while when the nanowire is in its topological superconducting phase, the presence of MBSs manifests in the reduction of linear conductance peak to $e^2/2h$. In addition, it was shown that the linear conductance through a QD, coupled to two MZMs confined in the opposite sites of a TSNW, where the QD-MZM ring structure is threaded by a magnetic field, presents a 2π-periodicity in magnetic flux phase when the Majorana wavefunctions do not overlap [2].

Over the last few years, various QD-TSNW structures have been theoretically proposed in order to detect the fingerprints of Majoranas via transport properties, such as conductance [4–24], current noise [4, 25–34], thermal conductance [35–45], optical schemes [43, 46–48], the Josephson current [49–53], and Fano resonance [54–67]. The photon-assisted tunneling in QD-MBS devices have been investigated both theoretically [68–72] and experimentally [73]. Zeng and co-workers [59] theoretically analyzed the transport properties of a dot-TSNW ring structure threaded by magnetic flux. They observed that for an asymmetrically biased QD the differential conductance exhibits a 2π-periodicity with the variation of magnetic flux phase for unhybridized and 4π for hybridized MBSs. It has been demonstrated that the transport in a complex T-shaped double-QD junction coupled to a superconductor (SC) lead can be corresponded to the transport in a dot-Majorana ring structure when fine-tuning the setups parameters [65, 74]. Other theoretical works on probe of shot noise in toroidal carbon nanotubes [75] or in monolayer graphene QDs [76] coupled to Majorana nanowires have been considered in order to detect the presence of MZMs.

In recent years, it has been shown that entanglement can be reached between two QDs connected to a pair of Majoranas confined in the opposite sites of a TSNW due to their nonlocal quantum nature [77–79]. Numerous works predicted that the nonlocality of MZMs results in the appearance of nonlocal processes, such as the electron tunneling (ET) as well as the crossed Andreev reflection (CAR) [79–83]. Beside the nonlocal processes, the local Andreev reflection (LAR) is also present in such devices [80–83]. It has been shown that the CAR and LAR processes can be controlled via Cooper-pair splitters when varying the gate voltages. Such Cooper-pair splitters have a Y-shaped structure in which two normal leads couple to a SC electrode through QDs [84]. Theoretical studies have demonstrated that artificial Kitaev chains can be realized using proximitized QDs arranged in arrays and connected via superconducting regions [85–90]. Nowadays, advances in nanofabrication technology have enabled the experimental realization of such minimal Kitaev chains [91–94]. Recently, researchers have made a great effort to the theoretical exploration of Andreev transport in various devices, such as in multi-QD-MZMs junctions [95–102]. All these studies exhibit complicated transport behavior and indicate that such devices can provide an ideal platform to detect the signature of MBSs [102].

Over the past decade, a plethora of experimental realizations on the detection of Majorana signatures via transport measurements have been reported by using devices based on semiconductor-SC nanowire junctions [103–122] or different QD-TSNW configurations [123–125]. However, these transport characteristics measurements can show significant deviations from theoretically predicted results, which may be due to the presence of decoherence in such devices. Decoherence appears when the

TSNW is connected to an environmental bath [126–129], which can be bosonic [126, 127, 130–136] or fermionic [130, 137, 138]. In such manner, decoherence can be caused by electromagnetic fields [134, 139], charge and thermal fluctuations in gate voltage [126, 134, 136, 140].

Recently, the condensed matter community paid a great attention to the investigation of the influence of optical phonons on transport characteristics of QD nanodevices. Hence, most of theoretical works focused on the exploration of phonon-assisted Andreev transport in different junctions such that in normal lead-SC [141], normal-QD-SC [142–148], or ferromagnet-QD-SC structures [149–152]. The phonon-assisted Andreev tunneling has been experimentally confirmed in a carbon nanotube QD attached to Nb SC and normal leads [153]. In particular, a couple of theoretical studies have addressed the phonon-assisted transport mechanisms at MZMs [154–158]. The transport properties of a dot-MZM setup with EPI were theoretically investigated before and yielded diverging findings about the behavior of linear conductance at zero temperature. The latter was supposedly demonstrated either as independent [156] on the electron-phonon coupling strength or dependent [155]. It was shown that the linear conductance can be modified by varying the electron-phonon coupling strength or the QD-MBS coupling values at zero temperature. However, at finite temperatures, the presence of EPI leads to the suppression of zero-bias differential conductance magnitude. Furthermore, the connection between the heat generation and electrical current in QD-TSNW systems in the presence of optical phonons was studied [157, 158].

As we have seen above, the theoretical study of devices for detection of MZMs still needs additional effort in the scientific community. The aim of this chapter is to propose a realistic, experimentally feasibly setup, that can probe MBSs via transport measurements. We show that the characteristic signatures expected from MZMs are significantly modified in the presence of EPI. We theoretically investigate the transport properties of a QD coupled to two MZMs, located at the opposite ends of a Majorana nanowire. The dot-TSNW junction forms a loop structure which is threaded by a magnetic flux in the presence of an optical phonon mode. The tunable magnetic flux allows one to manipulate the electronic transport by modulating the magnetic flux. In particular, we study the phonon-assisted tunneling current behavior in a system where a single QD is attached to one MZM located at one of the edges of a TSNW. As we have seen above, this problem is relevant because the quantum system is generally connected to an environment which can thus induce decoherence effects on MBSs-based quantum computation. We model the QD-Majorana system with a Holstein-type Hamiltonian and treat the EPI by applying a canonical transformation within the nonequilibrium Green's function (NGF) formalism. A brief discussion on transport processes that take place in such devices is also presented. Details on calculations regarding the currents, linear and differential conductances in both absence and presence of EPI can be found in our studies [159, 160]. In the last part of this chapter, we summarize our results. Here, we just briefly outline the central findings of our investigations. In the case when the QD hybridizes with both MZMs located in opposite sites of a Majorana nanowire, we show that without EPI the linear conductance presents a 2π periodicity in magnetic flux phase when the QD energy is zero, regardless the values of Majorana overlap energy. We demonstrate that the 2π periodicity of linear conductance transforms to 4π when tuning the dot level or introducing EPI for hybridized MBSs. The zero-temperature linear conductance is immune to the changes in QD energy, finite values of dot-MZM couplings or the presence of EPI for nonoverlapping Majorana wavefunctions. The linear conductance at finite temperatures depends on QD energy, electron-phonon coupling strength and finite values of QD-MZM couplings even for degenerate Majoranas. We show that the periodicity of differential conductance in magnetic flux phase can be switched from 2π to 4π for overlapping Majoranas when varying the QD energy or EPI strength. We also demonstrate that in the absence of EPI the periodicity of differential and linear conductances remains unchanged for a possible asymmetry in dot-lead couplings which would help the experimental verification. The devices studied in this chapter can serve as a possible guide for Majorana probing experiments based on QDs, helping thus to enlarge our knowledge of topological quantum computation.

2.2 Theoretical Model and Analytical Results

2.2.1 Model Hamiltonian of the System

In this section, we introduce our QD-MZM setup in the presence of EPI. The system consists of a dot which is connected to two Majoranas hosted by the ends of a TSNW. The QD-TSNW configuration forms a loop structure that is threaded by an external magnetic flux Φ, as it is illustrated in Figure 2.1. Thus, the transport properties of the QD-TSNW nanojunction can be manipulated when tuning the magnetic flux. The normal leads allow to measure the current through the dot. To investigate the effect of the EPI on transport properties, we consider that the localized electron in the dot interacts with a single long-wave optical phonon mode. The Majoranas are attached to the QD in the Coulomb blockade regime, thus the dot can be represented as a single fermion level [1, 30]. Furthermore, in order to drive the semiconducting nanowire into the topological superconducting phase, the Zeeman energy $h = g_W \mu_B B/2$, induced by the applied magnetic field (B), should be larger than $h > \sqrt{\Delta^2 + \mu^2}$ where Δ and μ represent the s-wave SC proximity induced pairing and chemical potential in the nanowire, respectively (for more details, see Sect. 1.2.2). Here, g_W is the Landé factor, while μ_B represents the Bohr magneton. The Zeeman energy is considered to be the largest energy scale, such that it is larger than the bias voltage $|eV|$, system's temperature T, the QD-lead coupling Γ_α, or QD-MBS coupling strength $|\lambda_j|$. Consequently, it is enough to assume a single spinless level in the dot [2, 4]. The proposed system is modeled by the Hamiltonian [2, 10, 58, 59, 159, 161, 162]

$$\mathcal{H} = \mathcal{H}_{\text{Leads}} + \mathcal{H}_{\text{MBS}} + \mathcal{H}_{\text{ph}} + \mathcal{H}_{\text{QD}} + \mathcal{H}_{\text{T}}. \tag{2.1}$$

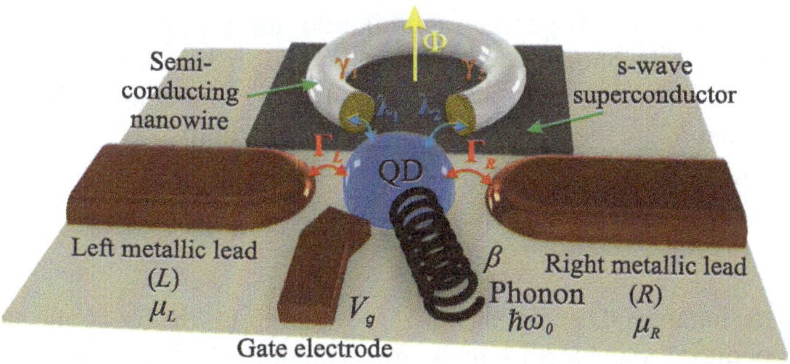

Fig. 2.1 Schematic representation of a QD attached to two MBSs, denoted by Majorana operators γ_1 and γ_2, via coupling strengths λ_1 and λ_2, hosted by the ends of a TSNW. The QD-TSNW setup forms a loop structure which is threaded by a tunable magnetic flux Φ. The QD is connected to left (L) and right (R) normal leads via coupling strengths Γ_L and Γ_R, respectively. The chemical potential in electrode $L(R)$ is denoted by $\mu_{L(R)}$. The QD level is tuned by the gate voltage V_g applied to the gate electrode. The QD electron interacts with a single long-wave optical phonon mode of frequency ω_0 via the electron-phonon coupling strength β. Reprinted with permission from [159]. Copyright 2022, American Physical Society

The first term of $\mathcal{H}$ in Eq. (2.1), $\mathcal{H}_{\text{Leads}}$, is the Hamiltonian of the normal leads and describes the motion of free electrons in both electrodes. It takes the form

$$\mathcal{H}_{\text{Leads}} = \sum_{\alpha,k} \varepsilon_{\alpha k}\, c_{\alpha k}^{\dagger} c_{\alpha k}, \tag{2.2}$$

where $c_{\alpha k}$ ($c_{\alpha k}^{\dagger}$) denotes the annihilation (creation) operator for a noninteracting electron with momentum k in the normal lead α [where $\alpha \equiv$ left (L), and right (R)]. Here, $\varepsilon_{\alpha k} = \varepsilon_k - \mu_\alpha$ correspond to single-particle energies measured from the chemical potential of lead α, μ_α. The last one is considered to be independent of the temperature. Furthermore, the electrodes are at the same $T_\alpha = T$ temperature.

The second term in Eq. (2.1), $\mathcal{H}_{\text{MBS}}$, is the MBS Hamiltonian which characterizes the coupling between the two Majoranas, hosted by the opposite ends of the TSNW loop. The Hamiltonian $\mathcal{H}_{\text{MBS}}$ is written as

$$\mathcal{H}_{\text{MBS}} = i\varepsilon_M \gamma_1 \gamma_2, \tag{2.3}$$

where $\varepsilon_M \propto e^{-L/\xi}$ denotes the overlap energy between Majorana wavefunctions that exponentially depends on the ratio of the TSNW loop length (L) and Majorana localization length (ξ). Here, γ_1 and γ_2 represent the Majorana operators corresponding to the two MZMs located at the opposite edges of a TSNW (see Sect. 1.2.2 for more details).

The next term in Eq. (2.1), $\mathcal{H}_{\text{ph}}$, represents the longitudinal optical phonon mode:

$$\mathcal{H}_{\text{ph}} = \hbar\omega_0 a^{\dagger} a, \tag{2.4}$$

where a and $a^\dagger$ denote the phonon annihilation and creation operators with $\hbar\omega_0$ being the energy of the phonon mode. The phonon distribution is determined by the Bose-Einstein function $N_{\text{ph}} = 1/(e^{\hbar\omega_0/k_B T} - 1)$. In the following, we set the reduced Planck's constant $\hbar$ and the Boltzmann constant k_B to unity.

In addition, $\mathcal{H}_{\text{QD}}$ in Eq. (2.1), describes the single-level QD:

$$\mathcal{H}_{\text{QD}} = \varepsilon_d d^\dagger d + \beta(a + a^\dagger) d^\dagger d, \tag{2.5}$$

where ε_d is the tunable QD energy, while d ($d^\dagger$) represents the fermionic annihilation (creation) operator for an electron in the dot. The second term of $\mathcal{H}_{\text{QD}}$ characterizes the interaction between the electron in QD and the optical phonon mode. Here, β symbolizes the electron-phonon coupling strength.

The last term in Eq. (2.1), $\mathcal{H}_{\text{T}}$, stands for the tunneling Hamiltonian and models the coupling of the dot to the Majoranas and to the normal leads:

$$\mathcal{H}_{\text{T}} = \left[(\lambda_1 d - \lambda_1^* d^\dagger)\gamma_1 + i(\lambda_2 d + \lambda_2^* d^\dagger)\gamma_2\right] + \sum_{\alpha,k} \left(V_{\alpha k} c_{\alpha k}^\dagger d + V_{\alpha k}^* d^\dagger c_{\alpha k}\right). \tag{2.6}$$

The first term in Eq. (2.6) represents the coupling between the QD electron and the two MZMs located at the sites of the Majorana nanowire. The QD connects to the MZMs via the coupling strengths λ_1 and λ_2. We assume that the couplings λ_1 and λ_2 are complex, $\lambda_1 = |\lambda_1| e^{i\phi/4}$ and $\lambda_2 = |\lambda_2| e^{-i\phi/4}$ [59]. Here, ϕ is the magnetic flux phase difference between the two QD-MBS couplings which is regulated by the threading magnetic flux Φ as $\phi = \pi\Phi/\Phi_0 = 2\arg(\lambda_1/\lambda_2)$, where $\Phi_0 = h/2e$ represents the magnetic flux quantum. The second term in $\mathcal{H}_{\text{T}}$ given by Eq. (2.6) describes the coupling between the localized electron in the dot and the free electrons in normal leads with the corresponding tunneling amplitude $V_{\alpha k}$.

In the following, we use the regular fermionic form of Majorana operators γ_1 and γ_2, which are defined as $\gamma_1 = (f^\dagger + f)/\sqrt{2}$ and $\gamma_2 = i(f^\dagger - f)/\sqrt{2}$, respectively (see also Sect. 1.2.2 for a general discussion). In this case, Eqs. (2.3) and (2.6) are rewritten as

$$\mathcal{H}_{\text{MBS}} = \varepsilon_M \left(f^\dagger f - \frac{1}{2}\right), \tag{2.7}$$

and

$$\begin{aligned}\mathcal{H}_{\text{T}} &= \frac{1}{\sqrt{2}}[(\lambda_1 + \lambda_2) df + (\lambda_1 - \lambda_2) df^\dagger - (\lambda_1^* - \lambda_2^*) d^\dagger f \\ &\quad - (\lambda_1^* + \lambda_2^*) d^\dagger f^\dagger] + \sum_{\alpha,k} \left(V_{\alpha k} c_{\alpha k}^\dagger d + V_{\alpha k}^* d^\dagger c_{\alpha k}\right).\end{aligned} \tag{2.8}$$

2.2.2 Current Formulas

In the framework of NGF formalism, the current flowing from the α normal lead to the QD is given by the relation [147, 163–170]:

$$I_\alpha = \frac{ie}{h}\int d\varepsilon \{\mathbf{\Gamma}_\alpha(\varepsilon)[\mathbf{f}_\alpha(\varepsilon)[\mathbf{G}_d^r(\varepsilon) - \mathbf{G}_d^a(\varepsilon)] + \mathbf{G}_d^<(\varepsilon)]\}_{11}, \tag{2.9}$$

where $\mathbf{G}_d^{r(a)}(\varepsilon)$ and $\mathbf{G}_d^{<(>)}(\varepsilon)$ are the retarded (advanced) and lesser (greater) Green's function matrices, respectively, of the dot represented in the Nambu space [171]:

$$\mathbf{G}_d^{r(a)}(\varepsilon) = \begin{pmatrix} G_{d11}^{r(a)}(\varepsilon) & G_{d12}^{r(a)}(\varepsilon) \\ G_{d21}^{r(a)}(\varepsilon) & G_{d22}^{r(a)}(\varepsilon) \end{pmatrix}, \quad \mathbf{G}_d^{<(>)}(\varepsilon) = \begin{pmatrix} G_{d11}^{<(>)}(\varepsilon) & G_{d12}^{<(>)}(\varepsilon) \\ G_{d21}^{<(>)}(\varepsilon) & G_{d22}^{<(>)}(\varepsilon) \end{pmatrix}, \tag{2.10}$$

with the relation $\mathbf{G}_d^r(\varepsilon) - \mathbf{G}_d^a(\varepsilon) = \mathbf{G}_d^>(\varepsilon) - \mathbf{G}_d^<(\varepsilon)$ [172–174]. The non-diagonal matrix elements of the retarded (advanced) Green's function, $G_{d12}^{r(a)}(\varepsilon)$ and $G_{d21}^{r(a)}(\varepsilon)$, represent the anomalous Green's functions where the subscript notations 1 and 2 refer to the electron and hole sector. The dot-normal lead coupling matrix reads [76, 152]

$$\mathbf{\Gamma}_\alpha(\varepsilon) = \begin{pmatrix} \Gamma_\alpha^e & 0 \\ 0 & \Gamma_\alpha^h \end{pmatrix}, \tag{2.11}$$

which elements in the wide-band limit [166], considered here, stand for $\Gamma_\alpha^{e(h)} = 2\pi\sum_k |V_{\alpha k}|^2\delta(\varepsilon \mp \varepsilon_{\alpha k})$ the coupling strength between the dot and normal lead α for electrons (holes). The elements of Fermi-Dirac distribution matrix for electrode α, $\mathbf{f}_\alpha(\varepsilon)$, are given as [145, 152, 168, 169]

$$\mathbf{f}_\alpha(\varepsilon) = \begin{pmatrix} f_\alpha^e(\varepsilon) & 0 \\ 0 & f_\alpha^h(\varepsilon) \end{pmatrix}, \tag{2.12}$$

with $f_\alpha^e(\varepsilon) = 1/[e^{(\varepsilon-\mu_\alpha)/T} + 1]$ and $f_\alpha^h(\varepsilon) = 1 - f_\alpha^e(-\varepsilon)$ being the Fermi-Dirac distribution function for electrons and holes, respectively. Furthermore, we define the spectral function of the QD, $\mathcal{A}_d(\varepsilon)$, as follows [162, 173]:

$$\mathcal{A}_d(\varepsilon) = \frac{i}{2}[\mathbf{G}_d^>(\varepsilon) - \mathbf{G}_d^<(\varepsilon)]_{11} = \frac{i}{2}[\mathbf{G}_d^r(\varepsilon) - \mathbf{G}_d^a(\varepsilon)]_{11}. \tag{2.13}$$

Now, substituting Eqs. (2.10)- (2.12) into Eq. (2.9), the current formula for lead α can be expressed in the following form:

$$I_\alpha = \frac{ie}{h}\int d\varepsilon\, \Gamma_\alpha^e[f_\alpha^e(\varepsilon)G_{d11}^>(\varepsilon) - [f_\alpha^e(\varepsilon) - 1]G_{d11}^<(\varepsilon)]. \tag{2.14}$$

We can determine the current I_S, flowing from the SC to the QD, with the help of Kirchhoff's law as $I_S + I_L + I_R = 0$ [168]. The current I_α can be further decomposed by applying the Keldysh equation $\mathbf{G}_d^{<(>)} = \mathbf{G}_d^r \Sigma^{<(>)} \mathbf{G}_d^a$. The lesser (greater) self-energy matrix $\Sigma^{<(>)}$ in the relatively weak β limit considered here is approximated as $\Sigma^{<(>)} \approx \Sigma_0^{<(>)}$ [161] where the noninteracting self-energies $\Sigma_0^{<(>)}(\varepsilon)$ are given by [76]

$$\Sigma_0^{<(>)}(\varepsilon) = \begin{pmatrix} \Sigma_{11}^{<(>)}(\varepsilon) & 0 \\ 0 & \Sigma_{22}^{<(>)}(\varepsilon) \end{pmatrix}, \tag{2.15}$$

with electron $\Sigma_{11}^{<}(\varepsilon) = i \sum_\alpha \Gamma_\alpha^e f_\alpha^e(\varepsilon)$, $\Sigma_{11}^{>}(\varepsilon) = i \sum_\alpha \Gamma_\alpha^e (f_\alpha^e(\varepsilon) - 1)$, and hole self-energies $\Sigma_{22}^{<}(\varepsilon) = i \sum_\alpha \Gamma_\alpha^h f_\alpha^h(\varepsilon)$, $\Sigma_{22}^{>}(\varepsilon) = i \sum_\alpha \Gamma_\alpha^h (f_\alpha^h(\varepsilon) - 1)$. In addition, our interest is restricted to an energy window located around the Majorana zero-energy, smaller than the proximity induced gap, we can consider the subgap regime, i.e., $|eV| < \Delta$, when the current flowing from the normal lead α, given by the relation (2.14), can be decomposed as

$$I_\alpha = I_\alpha^{ET} + I_\alpha^{LAR} + I_\alpha^{CAR}, \tag{2.16}$$

$$I_\alpha^{ET} = \frac{e}{h} \int d\varepsilon \, T_{\alpha\alpha'}^{ee}(\varepsilon)[f_\alpha^e(\varepsilon) - f_{\alpha'}^e(\varepsilon)], \tag{2.17}$$

$$I_\alpha^{LAR} = \frac{e}{h} \int d\varepsilon \, T_{\alpha\alpha}^{eh}(\varepsilon)[f_\alpha^e(\varepsilon) - f_\alpha^h(\varepsilon)], \tag{2.18}$$

and

$$I_\alpha^{CAR} = \frac{e}{h} \int d\varepsilon \, T_{\alpha\alpha'}^{eh}(\varepsilon)[f_\alpha^e(\varepsilon) - f_{\alpha'}^h(\varepsilon)], \tag{2.19}$$

where I_α^{ET}, I_α^{LAR}, and I_α^{CAR} are the currents generated in the ET, LAR, and CAR processes, respectively. The ET process manifests in the transmission of an electron from one electrode to the other one without annihilating/creating Cooper pairs in the superconducting material. The LAR process refers to the tunneling of one electron/hole from one electrode into the SC via one MZM after that it is reflected as a hole/electron in the same electrode by injecting a Cooper pair into the superconducting material. In a CAR process, a pair of MZMs connects to two normal electrodes in which one electron/hole from one electrode tunnels into the superconducting material via one MZM and then it tunnels out as a hole/electron via the other MZM at the other electrode by splitting thus a Cooper pair over the electrodes [25, 80–83, 175]. The notations $T_{\alpha\alpha'}^{ee}(\varepsilon) = \Gamma_\alpha^e \Gamma_{\alpha'}^e |G_{d11}^r(\varepsilon)|^2$, $T_{\alpha\alpha}^{eh}(\varepsilon) = \Gamma_\alpha^e \Gamma_\alpha^h |G_{d12}^r(\varepsilon)|^2$, and $T_{\alpha\alpha'}^{eh}(\varepsilon) = \Gamma_\alpha^e \Gamma_{\alpha'}^h |G_{d12}^r(\varepsilon)|^2$ in current formulas (2.17)–(2.19) represent the ET, LAR, and CAR transmission probabilities. In the following, we assume electron-hole symmetry in QD-lead couplings, i.e., $\Gamma_\alpha^e = \Gamma_\alpha^h = \Gamma_\alpha$ [10] and consider symmetrically coupled QD systems which is a broadly used approximation in the literature, thus, $\Gamma_L = \Gamma_R = \Gamma$. We also consider a grounded SC, i.e., $\mu_S = 0$, and apply an opposite bias voltage between the normal leads as $\mu_L = -\mu_R = eV/2$ which choice of chemical potentials leads to the identity $f_\alpha^e(\varepsilon) = f_{\alpha'}^h(\varepsilon)$. In such a case, only the ET and LAR processes occur, while the CAR processes are inhibited [63]. Using the Kirchhoff's law and Eqs. (2.16)–(2.19), the current flowing from the SC becomes $I_S = 0$ because $I_L = -I_R$. Thus, by employing Eq. (2.14), the current through the dot can be symmetrized as $I = (I_L - I_R)/2$,

$$I = \frac{e}{h} \int d\varepsilon \, \Gamma \mathcal{A}_d(\varepsilon) \big[f_L^e(\varepsilon) - f_R^e(\varepsilon) \big]. \tag{2.20}$$

In summary, the evaluation of the spectral function $\mathcal{A}_d(\varepsilon)$ in Eq. (2.13), the current I in the case of $\Gamma_L = \Gamma_R = \Gamma$ and $\mu_L = -\mu_R = eV/2$, given by Eq. (2.20), can be performed via the determination of component 11 of the lesser and greater Green's function matrices, $G_{\tilde{d}11}^<(\varepsilon)$ and $G_{\tilde{d}11}^>(\varepsilon)$, respectively. To calculate $G_{\tilde{d}11}^{<(>)}(\varepsilon)$ in the presence of EPI, we decouple the EPI in the total Hamiltonian given by Eq. (2.1) into two subsystems, into an electron and a phonon part. This method, called *canonical transformation*, is presented in Sect. 2.2.3.

2.2.3 Canonical Transformation

In this section, we employ a canonical transformation on the Hamiltonian of the system, $\mathcal{H}$, given by Eq. (2.1), in order to eliminate the electron-phonon coupling term. By doing so, the lesser and greater Green's functions, $\mathbf{G}_{\tilde{d}}^<(\varepsilon)$ and $\mathbf{G}_{\tilde{d}}^>(\varepsilon)$, can be determined in the presence of EPI. The canonical transformation consists of the application of operator $S = (\beta/\omega_0)d^\dagger d(a^\dagger - a)$ on $\mathcal{H}$, which transforms it into a new Hamiltonian as $\bar{\mathcal{H}} = e^S \mathcal{H} e^{-S}$ [161, 162, 176, 177]. The transformed Hamiltonian is now decoupled into a phonon and an electron part as $\bar{\mathcal{H}} = \mathcal{H}_{\rm ph} + \bar{\mathcal{H}}_{\rm el}$ where the phonon term, $\mathcal{H}_{\rm ph}$, remains unchanged, and it is given by the relation (2.4). The electron term of $\bar{\mathcal{H}}$, i.e., $\bar{\mathcal{H}}_{\rm el}$, takes the form

$$\bar{\mathcal{H}}_{\rm el} = \mathcal{H}_{\rm Leads} + \mathcal{H}_{\rm MBS} + \tilde{\mathcal{H}}_{\rm QD} + \bar{\mathcal{H}}_{\rm T}, \tag{2.21}$$

where the transformed tunneling Hamiltonian reads

$$\begin{aligned} \bar{\mathcal{H}}_{\rm T} &= \frac{1}{\sqrt{2}}[(\tilde{\lambda}_1 + \tilde{\lambda}_2)df + (\tilde{\lambda}_1 - \tilde{\lambda}_2)df^\dagger - (\tilde{\lambda}_1^* - \tilde{\lambda}_2^*)d^\dagger f \\ &\quad -(\tilde{\lambda}_1^* + \tilde{\lambda}_2^*)d^\dagger f^\dagger] + \sum_{\alpha,k} \big(\tilde{V}_{\alpha k} c_{\alpha k}^\dagger d + \tilde{V}_{\alpha k}^* d^\dagger c_{\alpha k}\big). \end{aligned} \tag{2.22}$$

The new QD Hamiltonian is defined as $\tilde{\mathcal{H}}_{\rm QD} = \tilde{\varepsilon}_d d^\dagger d$ with the renormalized dot energy $\tilde{\varepsilon}_d = \varepsilon_d - g\omega_0$, where $g = (\beta/\omega_0)^2$ represents a dimensionless EPI coupling parameter. We can observe that the transformed Hamiltonian $\bar{\mathcal{H}}_{\rm T}$ in Eq. (2.22) includes also the phonon operators via the new renormalized QD-lead α tunneling amplitude, $\tilde{V}_{\alpha k} = V_{\alpha k} X$, and the renormalized coupling strength between the dot and the jth MBS, $\tilde{\lambda}_j = \lambda_j X$, where $X = \exp(-\frac{\beta}{\omega_0}(a^\dagger - a))$. To eliminate the dependence of $\tilde{V}_{\alpha k}$ and $\tilde{\lambda}_j$ on phonon operators via X, we replace the operator X by its expectation value in thermal equilibrium, as $X \approx \langle X \rangle = \exp(-g(N_{\rm ph} + \frac{1}{2}))$ [162, 177]. This leads to $\tilde{V}_{\alpha k} \approx V_{\alpha k}\langle X \rangle$ and $\tilde{\lambda}_j \approx \lambda_j \langle X \rangle$. Now, the Hamiltonian $\bar{\mathcal{H}}_{\rm el}$, given by Eq. (2.21), is decoupled from the a and $a^\dagger$ phonon operators. This broadly employed approximation in literature is reasonable if $\beta \ll \min(V_{\alpha k}, \lambda_j, \Delta)$ or $V_{\alpha k}, \lambda_j \ll \min(\beta, \Delta)$, which yields that β, or $V_{\alpha k}$ and λ_j represent the smallest energy scale in the system. Note that this approximation is exact when $\beta = 0$ or $V_{\alpha k} = \lambda_j = 0$ [142, 154, 162, 177]. Thus, the greater $[\mathbf{G}_d^>(t)]$ and lesser $[\mathbf{G}_d^<(t)]$ Green's functions can be separated into two terms:

$$\begin{aligned}
\mathbf{G}_d^>(t) &= -i\langle d(t) d^\dagger \rangle \\
&= -i\langle e^{i\bar{\mathcal{H}}_{\rm el}t} d e^{-i\bar{\mathcal{H}}_{\rm el}t} d^\dagger \rangle_{\rm el} \langle e^{i\mathcal{H}_{\rm ph}t} X e^{-i\mathcal{H}_{\rm ph}t} X^\dagger \rangle_{\rm ph} \\
&= \tilde{\mathbf{G}}_d^>(t) \langle X(t) X^\dagger \rangle_{\rm ph}, \qquad (2.23) \\
\mathbf{G}_d^<(t) &= i\langle d^\dagger d(t) \rangle = \tilde{\mathbf{G}}_d^<(t) \langle X^\dagger X(t) \rangle_{\rm ph}, \qquad (2.24)
\end{aligned}$$

where $\tilde{\mathbf{G}}_d^>(t)$ and $\tilde{\mathbf{G}}_d^<(t)$ stand for the dressed greater and lesser Green's functions of an electron governed by the Hamiltonian $\bar{\mathcal{H}}_{\rm el}$. The phonon terms are given by $\langle X(t) X^\dagger \rangle_{\rm ph} = e^{-\Omega(t)}$ and $\langle X^\dagger X(t) \rangle_{\rm ph} = e^{-\Omega(-t)}$ with $\Omega(t) = g[N_{\rm ph}(1 - e^{i\omega_0 t}) + (N_{\rm ph} + 1)(1 - e^{-i\omega_0 t})]$ [162, 177]. Notice that the approximation $\langle X^\dagger X(t) \rangle_{\rm ph} = \langle X(t) X^\dagger \rangle_{\rm ph}$, employed in previous studies [161, 162, 178, 179], that directly decouples $\mathbf{G}_d^r(t)$, holds only at higher temperatures, where $N_{\rm ph}$ is much greater than one. At low temperatures where the phonon number is $N_{\rm ph} \sim \mathcal{O}(1)$, the above approximation, i.e., $\langle X^\dagger X(t) \rangle_{\rm ph} = \langle X(t) X^\dagger \rangle_{\rm ph}$, is no longer appropriate [162].

The phonon parts, $\langle X(t) X^\dagger \rangle_{\rm ph}$ and $\langle X^\dagger X(t) \rangle_{\rm ph}$, can be determined by applying the identity $e^{-\Omega(\mp t)} = \sum_{p=-\infty}^{\infty} \mathcal{L}_p e^{\pm i p \omega_0 t}$, where $\mathcal{L}_p$ denotes the Franck-Condon factor. The coefficients $\mathcal{L}_p$ of integer index p depend on the temperature T and electron-phonon coupling strength β. At nonzero temperatures, the Franck-Condon factor $\mathcal{L}_p$ takes the form [162, 177]:

$$\mathcal{L}_p = e^{-g(2N_{\rm ph}+1)} e^{p\omega_0/(2T)} I_p\big(2g\sqrt{N_{\rm ph}(N_{\rm ph}+1)}\big), \qquad (2.25)$$

where $I_p(z)$ stands for the pth-order modified Bessel function of the first kind. The coefficients $\mathcal{L}_p$ given by Eq. (2.25), at zero temperature, reduce to

$$\mathcal{L}_p = \begin{cases} e^{-g} g^p / p!, & p \geq 0, \\ 0, & p < 0. \end{cases} \qquad (2.26)$$

Applying the above identity for phonon terms, the Fourier transform of the greater $[\mathbf{G}_d^>(t)]$ and lesser $[\mathbf{G}_d^<(t)]$ Green's functions, given by Eqs. (2.23) and (2.24), are expressed as

$$\mathbf{G}_d^{>}(\varepsilon) = \sum_{p=-\infty}^{\infty} \mathcal{L}_p \tilde{\mathbf{G}}_d^{>}(\varepsilon - p\omega_0),$$
$$\mathbf{G}_d^{<}(\varepsilon) = \sum_{p=-\infty}^{\infty} \mathcal{L}_p \tilde{\mathbf{G}}_d^{<}(\varepsilon + p\omega_0). \tag{2.27}$$

To determine the dressed lesser, $\tilde{\mathbf{G}}_d^{<}(\varepsilon + p\omega_0)$, and dressed greater, $\tilde{\mathbf{G}}_d^{>}(\varepsilon - p\omega_0)$, Green's functions, we apply the Keldysh equation, $\tilde{\boldsymbol{G}}_d^{<(>)} = \tilde{\boldsymbol{G}}_d^{r} \tilde{\boldsymbol{\Sigma}}^{<(>)} \tilde{\boldsymbol{G}}_d^{a}$, where the components of dressed lesser (greater) self-energy $\tilde{\boldsymbol{\Sigma}}^{<(>)}$, following relation (2.15), are the electron $\tilde{\Sigma}_{11}^{<} = i\sum_\alpha \tilde{\Gamma}_\alpha^e f_\alpha^e$, $\tilde{\Sigma}_{11}^{>} = i\sum_\alpha \tilde{\Gamma}_\alpha^e (f_\alpha^e - 1)$, and hole self-energies $\tilde{\Sigma}_{22}^{<} = i\sum_\alpha \tilde{\Gamma}_\alpha^h f_\alpha^h$, $\tilde{\Sigma}_{22}^{>} = i\sum_\alpha \tilde{\Gamma}_\alpha^h (f_\alpha^h - 1)$. The mixed self-energies take $\tilde{\Sigma}_{12}^{<(>)} = \tilde{\Sigma}_{21}^{<(>)} = 0$. Using the relations for dressed greater and lesser self-energies, the corresponding Green's functions read

$$\begin{aligned}\tilde{G}_{d11}^{>} &= \tilde{\Sigma}_{11}^{>}|\tilde{G}_{d11}^{r}|^2 + \tilde{\Sigma}_{22}^{>}|\tilde{G}_{d12}^{r}|^2,\\ \tilde{G}_{d11}^{<} &= \tilde{\Sigma}_{11}^{<}|\tilde{G}_{d11}^{r}|^2 + \tilde{\Sigma}_{22}^{<}|\tilde{G}_{d12}^{r}|^2.\end{aligned} \tag{2.28}$$

The retarded Green's functions of the QD, $\tilde{G}_{d11}^r(\varepsilon)$ and $\tilde{G}_{d12}^r(\varepsilon)$, respectively, are determined employing the equation of motion (EOM) technique [159, 180–182]. The EOM method allows us to analytically determine the single-particle Green's function. For quantum systems described by an Anderson-like model, the EOM method leads to a hierarchical infinite number of higher-order correlation functions that must be truncated. An adequate truncation of the correlation function hierarchy determines the physical effects that can be observed in the system [181]. The detailed calculations of the corresponding retarded Green's functions of the dot for arbitrary couplings $\tilde{\lambda}_1$ and $\tilde{\lambda}_2$, as well as for $\tilde{\Gamma}_L$ and $\tilde{\Gamma}_R$, are presented in Appendix. Here, we first give the final results for a setup illustrated in Fig. 2.1 in which the QD connects to both the ends of a TSNW where this Majorana loop is threaded by a magnetic flux Φ. In this case the QD-MBS couplings are $\tilde{\lambda}_1 = |\tilde{\lambda}_1|e^{i\phi/4}$ and $\tilde{\lambda}_2 = |\tilde{\lambda}_2|e^{-i\phi/4}$, leading to

$$\tilde{G}_{d11}^r(\varepsilon) = \Big\{(\varepsilon + \tilde{\varepsilon}_d + i\tilde{\Gamma}) - \Big[|\tilde{\lambda}_1|^2 + |\tilde{\lambda}_2|^2 + \frac{2\varepsilon_M}{\varepsilon}|\tilde{\lambda}_1\tilde{\lambda}_2|\cos\frac{\phi}{2}\Big]\mathcal{K}(\varepsilon)\Big\}\tilde{\mathcal{F}}(\varepsilon)^{-1}, \tag{2.29a}$$

$$\tilde{G}_{d12}^r(\varepsilon) = (|\tilde{\lambda}_2|^2 e^{i\phi/2} - |\tilde{\lambda}_1|^2 e^{-i\phi/2})\mathcal{K}(\varepsilon)\tilde{\mathcal{F}}(\varepsilon)^{-1}, \tag{2.29b}$$

$$\tilde{\mathcal{F}}(\varepsilon) = (\varepsilon - \tilde{\varepsilon}_d + i\tilde{\Gamma})(\varepsilon + \tilde{\varepsilon}_d + i\tilde{\Gamma}) - 2(\varepsilon + i\tilde{\Gamma})(|\tilde{\lambda}_1|^2 + |\tilde{\lambda}_2|^2)\mathcal{K}(\varepsilon) + \frac{4|\tilde{\lambda}_1\tilde{\lambda}_2|^2}{\varepsilon}\mathcal{K}(\varepsilon)\cos^2\frac{\phi}{2} + \frac{4\varepsilon_M}{\varepsilon}\tilde{\varepsilon}_d|\tilde{\lambda}_1\tilde{\lambda}_2|\mathcal{K}(\varepsilon)\cos\frac{\phi}{2}, \tag{2.29c}$$

where $\mathcal{K}(\varepsilon) = \varepsilon/(\varepsilon^2 - \varepsilon_M^2)$ and $\tilde{\Gamma} = \frac{1}{2}\sum_\alpha \tilde{\Gamma}_\alpha$ by assuming electron-hole symmetry in the QD-lead couplings, i.e., $\tilde{\Gamma}_\alpha^e = \tilde{\Gamma}_\alpha^h = \tilde{\Gamma}_\alpha$, as mentioned above. Note here that we can reach the same result for $\tilde{G}_{d11}^r(\varepsilon)$ as in Ref. [59] under mapping $\tilde{\Gamma} \to \Gamma/2$ with the exception of ε_M sign due to the distinct sign of γ_1 in the tunneling Hamiltonian (2.6). In addition, we can simply obtain the retarded Green's functions for a system studied below where the QD couples to one of the ends of a TSNW by assuming $|\lambda_1| = |\lambda|$ and $|\lambda_2| = 0$ with $\phi = 0$ in Eqs. (2.29a)–(2.29c). Thus, one finds

$$\tilde{G}^r_{d11}(\varepsilon) = \frac{\varepsilon + \tilde{\varepsilon}_d + i\tilde{\Gamma} - |\tilde{\lambda}|^2\mathcal{K}(\varepsilon)}{(\varepsilon - \tilde{\varepsilon}_d + i\tilde{\Gamma})(\varepsilon + \tilde{\varepsilon}_d + i\tilde{\Gamma}) - 2(\varepsilon + i\tilde{\Gamma})|\tilde{\lambda}|^2\mathcal{K}(\varepsilon)},$$
$$\tilde{G}^r_{d12}(\varepsilon) = \frac{-|\tilde{\lambda}|^2\mathcal{K}(\varepsilon)}{(\varepsilon - \tilde{\varepsilon}_d + i\tilde{\Gamma})(\varepsilon + \tilde{\varepsilon}_d + i\tilde{\Gamma}) - 2(\varepsilon + i\tilde{\Gamma})|\tilde{\lambda}|^2\mathcal{K}(\varepsilon)}. \tag{2.30}$$

Note here by setting $\varepsilon_M = 0$ in Eq. (2.30), the retarded Green's functions $\tilde{G}^r_{d11}(\varepsilon)$ and $\tilde{G}^r_{d12}(\varepsilon)$ reduce to those of Ref. [155]. Finally, the lesser (greater) Green's function of the QD, $\mathbf{G}_d^{<(>)}(\varepsilon)$, given by Eq. (2.27), can be determined with the help of Eqs. (2.28)–(2.30). The final formulas of the spectral function, linear and differential conductances, in the absence and presence of EPI, at zero and finite temperatures, can be found in our study [159].

2.3 Results and Discussion

In this section, we present and discuss the results for transport properties of the considered dot-MZMs system in both the absence and the presence of EPI within several experimentally relevant parameter regimes. As we also mentioned in Sect. 2.2, all the system parameters should be much smaller than the induced superconducting gap Δ which in typical experimental setups lies on the order of 250 μeV for TSNWs [103]. Such TSNWs can be realized from the deposition of an InSb semiconducting nanowire with strong Rashba-type spin-orbit interaction in the proximity of a s-wave SC in the presence of a magnetic field (for more details see Sect. 1.2.2). The values of QD-leads Γ and QD-MBS $|\lambda_j|$ couplings are on the order of a few μeV in experiments [4, 183]. In calculation, we measure all energies in units of Γ. Therefore, the EPI-related calculations are done in the limit where the symmetrical dot-lead couplings Γ and dot-MZM couplings $|\lambda_j|$ are weaker than the electron-phonon coupling strength β (see also Sect. 2.2.3). The value of electron-phonon coupling strength, with a few exceptions, is fixed as $\beta = 2.5\,\Gamma$. We consider that the energy of longitudinal optical phonons is on the order of a few tens of μeV [142, 184, 185] being thus much smaller than Δ. In the following calculations, the phonon energy is set as $\omega_0 = 5\,\Gamma$.

In such systems, where a dot couples to leads and MBSs once, different paths are present for charge carriers passing from one electrode to the other one through the QD. Paths that involve Andreev reflection at the interface of a TSNW can destructively and constructively interfere with those that travel through the QD, and thus they produce a Fano lineshape in transport spectrum [59, 61, 65]. The Fano structure emerges when the QD connects to the side-coupled MZMs, and thus electrons can scatter resonantly on those. Such kind of fractional Fano-type resonances can provide a clear evidence for the Majorana fingerprints in QDs side-attached to edges of TSNWs [60].

2.3.1 Results in the Absence of Electron-Phonon Interaction

As already stated above, we first explore the spectral function $\mathcal{A}_d(\varepsilon)$, and the ET and LAR processes contributing to the total differential and linear conductance when the QD is oppositely biased ($\mu_L = -\mu_R = eV/2$), in the absence of EPI.

In the following, instead of the QD spectral function $\mathcal{A}_d(\varepsilon)$, to simplify the discussion, we use the normalized spectral function $\Gamma\mathcal{A}_d(\varepsilon)$ that in the absence of EPI can be written as $\Gamma\mathcal{A}_d(\varepsilon) = -\Gamma \mathrm{Im} G^r_{d11}(\varepsilon) = \mathcal{T}^{ee}_{LR}(\varepsilon) + \mathcal{T}^{eh}_{LL}(\varepsilon)$. Therefore, in the approximation scheme applied here within the EOM technique, the dot retarded Green's functions $G^r_{d11}(\varepsilon)$ and $G^r_{d12}(\varepsilon)$, respectively, do not depend on bias voltage and temperature in the absence of EPI. The resulting normalized spectral function is also bias- and temperature-independent.

To better understand the spectral function in the no EPI case, we give the poles of the retarded Green's function $G^r_{d11}(\varepsilon)$ in the limit $\Gamma \to 0$ [186] which determine the location of maxima emerging in $\mathcal{A}_d(\varepsilon)$. The poles of $G^r_{d11}(\varepsilon)$, in general, when the QD couples to both the ends of the TSNW with couplings $\lambda_1 = |\lambda_1|e^{i\phi/4}$ and $\lambda_2 = |\lambda_2|e^{-i\phi/4}$ are expressed as

$$2\varepsilon^2 \approx 2(|\lambda_1|^2 + |\lambda_2|^2) + \varepsilon_d^2 + \varepsilon_M^2 \pm \sqrt{[2(|\lambda_1|^2+|\lambda_2|^2)+\varepsilon_d^2+\varepsilon_M^2]^2 - 4\{\varepsilon_d^2\varepsilon_M^2 + 4|\lambda_1||\lambda_2|\cos(\phi/2)[|\lambda_1||\lambda_2|\cos(\phi/2)+\varepsilon_d\varepsilon_M]\}}. \tag{2.31}$$

We first plot in Fig. 2.2 the normalized spectral function of the dot $\Gamma\mathcal{A}_d(\varepsilon)$ in the case when the QD is symmetrically attached to the electrodes and also to both the ends of an adequately long TSNW with vanishing Majorana overlap energy ($\varepsilon_M = 0$), when the symmetrical dot-MZM coupling $|\lambda_j| = |\lambda|$ and the magnetic flux phase ϕ are varied at dot energy level $\varepsilon_d = 0$. Note that the location of peaks appearing in $\Gamma\mathcal{A}_d(\varepsilon)$ are determined in Eq. (2.31) by considering $\varepsilon_M = \varepsilon_d = 0$ and $|\lambda_1| = |\lambda_2| = |\lambda|$. We observe that at small values of $|\lambda|$, the normalized spectral function $\Gamma\mathcal{A}_d(\varepsilon)$ contains two maxima at energies $\varepsilon \approx \pm\sqrt{2}|\lambda|\sqrt{1 + |\sin(\phi/2)|}$ which shift to the higher values of energy $|\varepsilon|$ with a further enhancement in $|\lambda|$. In addition, the zero-energy normalized spectral function $\Gamma\mathcal{A}_d(0)$ vanishes for nearly any magnetic phase value ϕ, except for the values $\phi = (2n+1)\pi$ with $n = 0, \pm1, \pm2, \ldots$ which results in $\Gamma\mathcal{A}_d(0) = 1/2$. Moreover, when the symmetrical QD-MZM coupling reaches the value $|\lambda| \approx \Gamma/2$, the normalized spectral function $\Gamma\mathcal{A}_d(\varepsilon)$ shows a narrow flat region in the vicinity of zero-energy, i.e., $|\varepsilon| \lesssim \Gamma/3$, at magnetic flux phase $\phi = (2n+1)\pi$, as was before reported in the literature [10]. After $|\lambda|$ exceeds $|\lambda| \approx \Gamma/2$, the spectrum of dot normalized spectral function $\Gamma\mathcal{A}_d(\varepsilon)$ exhibits a

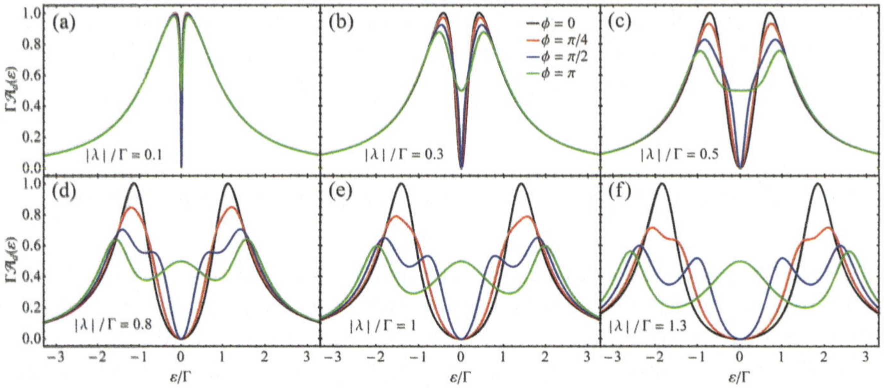

Fig. 2.2 The QD normalized spectral function $\Gamma\mathcal{A}_d(\varepsilon)$ for different magnetic flux phase ϕ values with unhybridized MBSs ($\varepsilon_M = 0\,\Gamma$), at dot energy $\varepsilon_d = 0\,\Gamma$, at symmetrical dot-MZM couplings $|\lambda_j| = |\lambda|$: **a** $|\lambda| = 0.1\,\Gamma$, **b** $|\lambda| = 0.3\,\Gamma$, **c** $|\lambda| = 0.5\,\Gamma$, **d** $|\lambda| = 0.8\,\Gamma$, **e** $|\lambda| = 1\,\Gamma$, and **f** $|\lambda| = 1.3\,\Gamma$. Reproduced with permission from [159]. Copyright 2022, American Physical Society

robust three-peak structure at $\phi = (2n+1)\pi$. As a consequence, for any finite symmetrical QD-MBS coupling values $|\lambda|$, at magnetic flux phases $\phi = (2n+1)\pi$, the zero-energy dot normalized spectral function is always $\Gamma\mathcal{A}_d(0) = 1/2$ for unhybridized Majoranas $\varepsilon_M = 0$, resulting thus in a linear conductance of $e^2/2h$. The $e^2/2h$ linear conductance value is a clear evidence of the Majorana fingerprints [2]. At $\phi = \pi/4$ and $\phi = \pi/2$, when $|\lambda|$ exceeds the value $|\lambda| > \Gamma/2$, the two-peak structure starts to evolve into a four-peak one. The side-peaks are located at $\varepsilon \approx \pm\sqrt{2}|\lambda|\sqrt{1+|\sin(\phi/2)|}$, while the position of new intermediate resonances is given by the relation $\varepsilon \approx \pm\sqrt{2}|\lambda|\sqrt{1-|\sin(\phi/2)|}$. In conclusion, the increasing symmetrical dot-MZM coupling $|\lambda|$ significantly modifies the peak structure of the dot normalized spectral function, but it has unremarkable influence on the spectrum of $\Gamma\mathcal{A}_d(\varepsilon)$ near zero-energy at magnetic flux phases $\phi \neq (2n+1)\pi$. Note also that $\Gamma\mathcal{A}_d(\varepsilon)$ is a 2π-periodic function of magnetic flux phase ϕ.

We next study the effect of QD energy ε_d and Majorana overlap energy ε_M on the QD normalized spectral function $\Gamma\mathcal{A}_d(\varepsilon)$ when the dot symmetrically couples to both MBSs of the TSNW. The results are shown in Fig. 2.3 for finite QD level $\varepsilon_d \neq 0$, overlap energies $\varepsilon_M \neq 0$ at different magnetic flux phase ϕ values with weak symmetrical dot-MZM couplings $|\lambda_1| = |\lambda_2|$. In experimental measurements, the QD level is tuned by the gate voltage via metallic gate electrodes. By shifting the QD level away from the Fermi energy, $\varepsilon_d \neq \varepsilon_F$, with $\varepsilon_F = 0$, the lineshape of normalized spectral function $\Gamma\mathcal{A}_d(\varepsilon)$ is visibly modified. At $\phi = 0$ [see Fig. 2.3a, b], the antiresonance point of $\Gamma\mathcal{A}_d(\varepsilon)$ reduces to zero and moves to the location $\varepsilon \approx -\varepsilon_M$ for any QD level ε_d values [64]. Furthermore, two maxima show up in the spectrum of $\Gamma\mathcal{A}_d(\varepsilon)$ of which the broadened peak is approximatively located at $\varepsilon \approx \varepsilon_d$. The narrow resonances are mostly associated to regular fermionic states coming from the MZMs. Additionally, in the special case when $\varepsilon_d = -\varepsilon_M$, the two

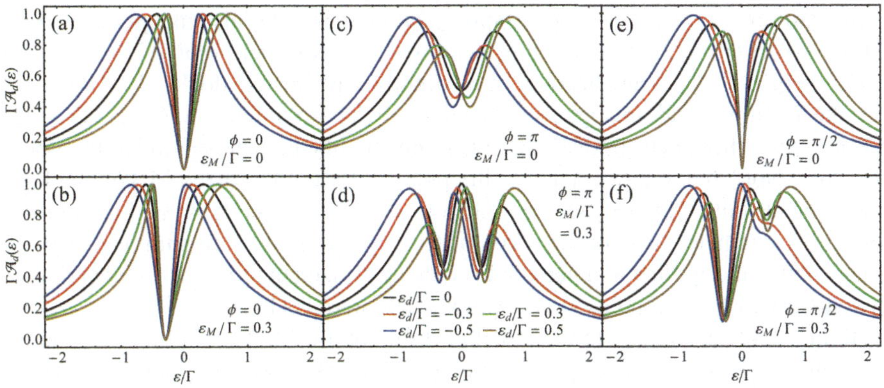

Fig. 2.3 The QD normalized spectral function $\Gamma\mathcal{A}_d(\varepsilon)$ for different values of QD energy ε_d, overlap energy ε_M and magnetic flux phase ϕ for symmetrical dot-MZM couplings $|\lambda_1| = |\lambda_2| = 0.3\,\Gamma$. The parameters are chosen as: **a** $\phi = 0$, $\varepsilon_M = 0\,\Gamma$; **b** $\phi = 0$, $\varepsilon_M = 0.3\,\Gamma$; **c** $\phi = \pi$, $\varepsilon_M = 0\,\Gamma$; **d** $\phi = \pi$, $\varepsilon_M = 0.3\,\Gamma$; **e** $\phi = \pi/2$, $\varepsilon_M = 0\,\Gamma$; and **f** $\phi = \pi/2$, $\varepsilon_M = 0.3\,\Gamma$. Reproduced with permission from [159]. Copyright 2022, American Physical Society

resonances in $\Gamma\mathcal{A}_d(\varepsilon)$ become identical, as previously reported in the literature [64]. At $\phi = \pi$ [see Fig. 2.3c, d], the two-peak structure of $\Gamma\mathcal{A}_d(\varepsilon)$ (appearing initially at $\varepsilon_M = 0$) transforms into a three-peak one with the increase of overlap energy ε_M confirming thus the presence of MZMs in the system [2, 69]. More importantly, in the case of degenerate Majoranas the zero-energy normalized spectral function $\Gamma\mathcal{A}_d(0)$ takes the 1/2 value regardless the dot level tuning, while the location of antiresonance point is affected by ε_d. Namely, it shifts to the positive or negative energy region for $\varepsilon_d > 0$ or $\varepsilon_d < 0$, respectively. In the case of hybridized MZMs, as the value of $|\varepsilon_d|$ increases, the magnitude of the intermediate resonance reduces from 1, while its location starts to shift away from $\varepsilon = 0$. Consequently, the main resonant peak shifts from the higher- to the lower-energy region when tuning the QD level from the empty ($\varepsilon_d > 0$) to the filled region ($\varepsilon_d < 0$). Note that beside the central resonance, two side-resonances emerge in the spectrum with magnitudes strongly influenced by the QD energy. The broadened peaks, which positions can be found with the relation (2.31), show an increased amplitude for large $|\varepsilon_d|$ values. The tuned QD energy level destructively affects the sharp peaks leading thus to resonances with smaller amplitude. Note that at the Fermi level, $\varepsilon_d = 0$, with weak dot-MZM coupling strengths considered here, the normalized spectral function $\Gamma\mathcal{A}_d(\varepsilon)$ displays two minima at $\varepsilon \approx \pm\varepsilon_M$. As the symmetrical QD-MZM couplings ($|\lambda_1| = |\lambda_2|$) increase, at any nonzero values of the overlap energy, the zero-energy normalized spectral function always displays $\Gamma\mathcal{A}_d(0) = 1$ instead of the 1/2 value at $\varepsilon_d = 0$ (not shown here). The increasing dot-MZM couplings leads to the minima of normalized spectral function move away from the energies $\varepsilon \approx \pm\varepsilon_M$. Also note that the full width at half-maximum of zero-energy resonance is proportional to the overlap energy ε_M. At $\phi = \pi/2$ [see Fig. 2.3e, f], the antiresonance point of the normalized spectral function is equal to zero at $\varepsilon = 0$ for nonoverlapping Majoranas and is also independent of the QD level ε_d. For finite ε_M values, the position of antiresonance point of $\Gamma\mathcal{A}_d(\varepsilon)$ shifts to the energy $\varepsilon \approx -\varepsilon_M$ and becomes now slightly dependent of dot level ε_d. We recall that the location of side-resonances is analytically approximated in Eq. (2.31). In the $\varepsilon_M \neq 0$ case, the spectrum of $\Gamma\mathcal{A}_d(\varepsilon)$ is more complicated, presenting a three-peak structure. Note that at any magnetic flux phase ϕ for unhybridized MBSs, the Fano lineshape at finite QD energy ε_d and the one at $-\varepsilon_d$ are clearly related by a mirror reflection symmetry around the zero-energy $\varepsilon = 0$ axis [64]. In summary, we can establish that the applied magnetic field could be utilized in flux-controlled operation in quantum computation based on topological materials [1, 10].

In Fig. 2.4, we investigate the effect of different dot-MZM couplings $|\lambda_j|$ on the spectrum of QD normalized spectral function $\Gamma\mathcal{A}_d(\varepsilon)$ at two values of the magnetic flux phase ϕ when the dot energy level is tuned $\varepsilon_d \neq 0$ for nonoverlapping Majoranas ($\varepsilon_M = 0$). Here, we study the features of $\Gamma\mathcal{A}_d(\varepsilon)$ by fixing one of the QD-MZM couplings $|\lambda_1| = 0.3\,\Gamma$ and changing $|\lambda_2|$. We can see that when the QD is not tuned ($\varepsilon_d = 0$) and also the magnetic flux is switched off ($\phi = 0$), two or four symmetrical resonances emerge in $\Gamma\mathcal{A}_d(\varepsilon)$ as a function of $|\lambda_2|$ values [see Fig. 2.4a]. When the QD connects only to one of the ends of a TSNW ($|\lambda_2| = 0$), the zero-energy normalized spectral function presents the expected $1/2$ value, $\Gamma\mathcal{A}_d(0) = 1/2$, which reduces to zero when $|\lambda_2|$ becomes finite. With the increase of $|\lambda_2|$, the resonances shift from the antiresonance point appearing at $\varepsilon = 0$ and their magnitude reaches a maximum at the value $|\lambda_1| = |\lambda_2|$. At $|\lambda_2| > |\lambda_1|$, the antiresonance valley becomes widened with a further increase in $|\lambda_2|$, as was previously reported [64]. This two-peak structure of $\Gamma\mathcal{A}_d(\varepsilon)$ is stable until the coupling $|\lambda_2|$ approaches the $\approx 0.8\,\Gamma$ value. A further enhancement in $|\lambda_2|$ leads to the spectrum of $\Gamma\mathcal{A}_d(\varepsilon)$ transforms into

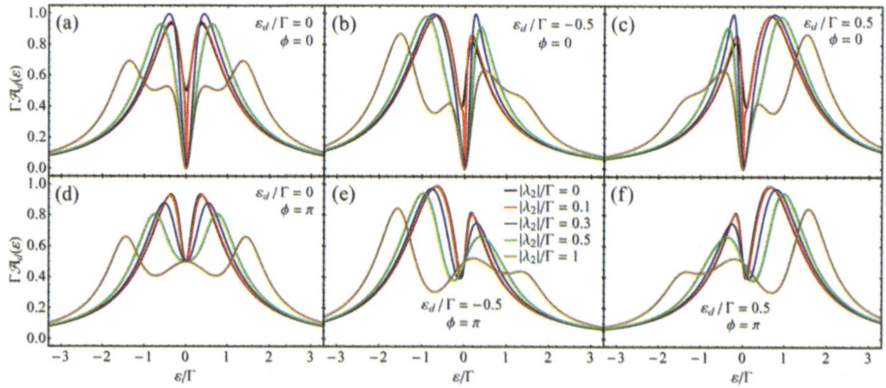

Fig. 2.4 The dot normalized spectral function $\Gamma\mathcal{A}_d(\varepsilon)$ for a tuned QD with $\varepsilon_M = 0\,\Gamma$ where the dot-MZM coupling $|\lambda_1|$ is fixed as $|\lambda_1| = 0.3\,\Gamma$ at different magnetic flux phase ϕ and dot-MZM coupling $|\lambda_2|$ values. The rest of parameters are: **a** $\phi = 0$, $\varepsilon_d = 0\,\Gamma$; **b** $\phi = 0$, $\varepsilon_d = -0.5\,\Gamma$; **c** $\phi = 0$, $\varepsilon_d = 0.5\,\Gamma$; **d** $\phi = \pi$, $\varepsilon_d = 0\,\Gamma$; **e** $\phi = \pi$, $\varepsilon_d = -0.5\,\Gamma$; and **f** $\phi = \pi$, $\varepsilon_d = 0.5\,\Gamma$. Reproduced with permission from [159]. Copyright 2022, American Physical Society

a four-resonance structure. Additionally, the minimum of $\Gamma \mathcal{A}_d(\varepsilon)$ at $|\lambda_2| = 0$ moves from the zero-energy position $\varepsilon = 0$ to the negative energies for dot level $\varepsilon_d < 0$, and to the positive energies for QD level $\varepsilon_d > 0$, respectively [see Fig. 2.4b, c]. Note that the location of the antiresonance point at zero-energy remains unchanged when the dot-MBS coupling $|\lambda_2|$ is finite, $|\lambda_2| \neq 0$, and the QD level is tuned, $\varepsilon_d \neq 0$. Accordingly to the findings from Fig. 2.3 where the dot symmetrically couples to both MBSs $|\lambda_1| = |\lambda_2|$, the magnitudes of broadened and sharp peaks associated with the MZMs, reach the value of 1 at $|\lambda_2| = |\lambda_1|$. When the coupling $|\lambda_2|$ exceeds the value of $|\lambda_1|$, the peaks' magnitude reduces. Furthermore, the two-peak structure of the normalized spectral function which is present at weak couplings $|\lambda_2|$ transforms into a four-peak structure at the value $|\lambda_2| \approx 1\,\Gamma$. When the threading magnetic flux is equal to $\phi = \pi$ [see Fig. 2.4d–f], then the zero-energy normalized spectral function for unhybridized MBSs ($\varepsilon_M = 0$) always becomes 1/2, i.e., $\Gamma \mathcal{A}_d(0) = 1/2$, at any values of the coupling $|\lambda_2|$ and dot level ε_d, respectively. In this case, the Majorana fingerprint is independent of the finite $|\lambda_j|$ values which can be interpreted as robust against disorder [95]. As in the $\phi = 0$ case, the spectrum of $\Gamma \mathcal{A}_d(\varepsilon)$ presents a two-peak structure at small values of $|\lambda_2|$, and a three-resonance structure can be observed in the spectrum at QD level $\varepsilon_d = 0$ for coupling $|\lambda_2| \gtrsim 0.7\,\Gamma$, while it is visible in $\Gamma \mathcal{A}_d(\varepsilon)$ at $\varepsilon_d \neq 0$ for coupling $|\lambda_2| \gtrsim 0.9\,\Gamma$. In addition, when the dot level is tuned ($\varepsilon_d \neq 0$), the antiresonance point shows the same behavior with the $\phi = 0$ case at coupling $|\lambda_2| = 0$. Note that in contrast to the findings from Fig. 2.4b, c, here when the QD level is tuned $\varepsilon_d \neq 0$, the location of antiresonance point for weak $|\lambda_2|$ couplings changes as in the $|\lambda_2| = 0$ case. Similarly, the location of peaks can be analytically approximated in Eq. (2.31) under considering $\varepsilon_M = 0$.

In conclusion, by summing the results of Figs. 2.3 and 2.4, we can establish that for $\varepsilon_M = 0$ the zero-energy normalized spectral function is $\mathcal{A}_d(0) = 1/2$, if either of the dot-MZM couplings is nonzero and the other one is zero at any flux phase ϕ, and $\mathcal{A}_d(0) = 1/2$ or $\mathcal{A}_d(0) = 0$ if both of the dot-MZM couplings exhibit finite values, $|\lambda_j| \neq 0$, at $\phi = (2n+1)\pi$ or $\phi \neq (2n+1)\pi$, respectively, independently of the changes in dot energy ε_d. This results in the linear conductance will be independent of dot level ε_d for unhybridized Majoranas at zero temperature (discussed in the following), which shows again the robustness of MZMs fingerprints [2, 5, 36].

We plot in Fig. 2.5a–c the linear conductance $\mathcal{G}$ and in Fig. 2.5d–f the differential conductance dI/dV at a fixed bias voltage eV value against the magnetic flux phase ϕ with different Majorana overlap energies ε_M at zero temperature. The dot hybridizes with both MBSs of symmetrical couplings $|\lambda_j| = |\lambda|$ at QD energy $\varepsilon_d = 0$. We see that the linear conductance $\mathcal{G}$ presents a 2π periodicity in ϕ with unhybridized ($\varepsilon_M = 0$) [2] and hybridized MZMs ($\varepsilon_M \neq 0$) where the maxima of peaks are located at magnetic flux phases $\phi = (2n+1)\pi$ (n is an integer). At magnetic flux phases $\phi \neq (2n+1)\pi$, the linear conductance equals $\mathcal{G} = 0$ for unhybridized MBSs ($\varepsilon_M = 0$), independent of the finite symmetrical dot-MZM couplings $|\lambda|$ values. The minima of linear conductance grow with the overlap energy ε_M. As the symmetrical dot-MBS coupling $|\lambda|$ increases, the linear conductance minima start to reduce and also tend to zero even at higher overlap energies ε_M, while the resonances get narrower. The

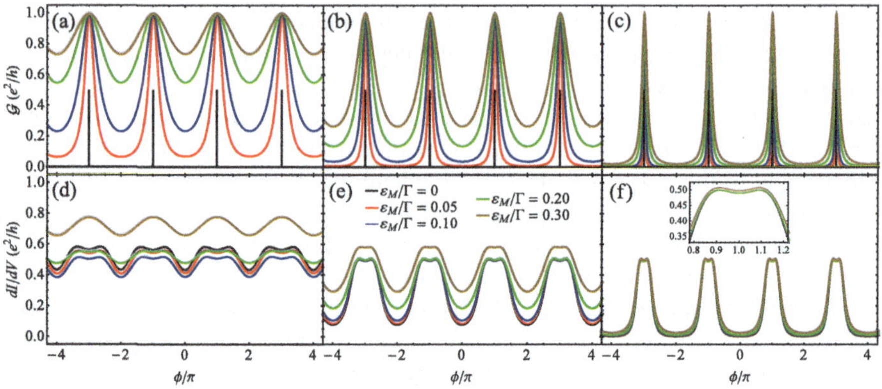

Fig. 2.5 The zero-temperature linear $\mathcal{G}$ **a–c** and zero-temperature differential conductance dI/dV at bias voltage $eV = 0.28\,\Gamma$ **d–f** versus magnetic flux phase ϕ at different values of the Majorana overlap energy ε_M for three values of the symmetrical dot-MZM couplings $|\lambda_j| = |\lambda|$ at QD energy $\varepsilon_d = 0\,\Gamma$. The parameters are chosen as: **a, d** $|\lambda| = 0.3\,\Gamma$, **b, e** $|\lambda| = 0.5\,\Gamma$, and **c, f** $|\lambda| = 1\,\Gamma$. The inset refers to the zoom at a peak of the differential conductance. Reprinted with permission from [159]. Copyright 2022, American Physical Society

overlap energy ε_M affects more obviously the width of conductance peaks when the dot weakly connects to the Majoranas. At zero temperature, the height of peaks is $e^2/2h$ for nonoverlapping and e^2/h for overlapping MZMs which is not influenced by the dot-MZM couplings $|\lambda_j| = |\lambda|$. The zero-temperature differential conductance dI/dV shows a 2π periodicity for nonoverlapping and overlapping Majoranas, with global minima emerging at magnetic flux phases $\phi = 2n\pi$, with n being an integer, similarly to the $\mathcal{G}$ case. At small values of ε_M, further local minima show up in dI/dV spectrum at magnetic flux phase values $\phi = (2n+1)\pi$. For stronger MBS-MBS coupling, the local minima of the dI/dV spectrum transform into maxima and all the differential conductance curve move upward [Fig. 2.5d]. We observe that the critical overlap energy value at which the local minima of dI/dV disappear, increases with the dot-MBS couplings. The increase of dot-MZM couplings $|\lambda|$ leads to the shift of the differential conductance curves to the position of that curve which corresponds to unhybridized MBSs ($\varepsilon_M = 0$). Therefore, the overlap energy ε_M has less impact on the differential conductance spectrum at larger values of the dot-MZM couplings $|\lambda|$ [Fig. 2.5e, f]. The differential conductance dI/dV has a plateau-like structure which peaks height approaches $e^2/2h$ and the width of maxima is also suppressed at stronger $|\lambda|$. Note that 2π and 4π periodicities of dI/dV as a function of magnetic flux phase ϕ have been communicated in the literature for QD-TSNW structures, with nonoverlapping and overlapping MZMs [59]. This discrepancy in reported results regarding the periodical behavior of differential conductance is attributed to the system biasing. To understand the influence of eV on the periodicity of dI/dV in magnetic flux phase ϕ, we introduce the following biasing: $\mu_L = qeV$ and $\mu_R = (q-1)eV$ such that the difference in chemical potentials fulfills $\mu_L - \mu_R = eV$ with the construction $0 \leq q \leq 1$. To do this, we focus on the zero-temperature differential conductance dI_L/dV calculated for current flowing from the left contact L to the QD, I_L. In this case, the zero-temperature differential conductance takes the form

$$\frac{dI_L}{dV} = \frac{e^2}{h}\left[qT_{LR}^{ee}(qeV) + (1-q)T_{LR}^{ee}((q-1)eV)\right] + \frac{e^2}{h}q\left[T_{LL}^{eh}(qeV) + T_{LL}^{eh}(-qeV)\right] + \frac{e^2}{h}\left[qT_{LR}^{eh}(qeV) + (q-1)T_{LR}^{eh}((1-q)eV)\right]. \tag{2.32}$$

We show in Fig. 2.6a, b the results for zero-temperature total differential conductance dI_L/dV against the magnetic flux phase ϕ for nonoverlapping ($\varepsilon_M = 0$) and overlapping ($\varepsilon_M \neq 0$) MBSs, respectively. Figure 2.6c illustrates the CAR component of the zero-temperature total differential conductance, i.e., dI_L^{CAR}/dV. The bias voltage eV is chosen as $eV = 0.28\,\Gamma$. The differential conductance dI_L/dV exhibits a 2π-periodic function in magnetic flux phase ϕ for unhybridized Majoranas and remains unchanged when altering the system biasing. When $\varepsilon_M \neq 0$, the periodicity of differential conductance can be 2π and 4π depending on how the biasing of the QD system is chosen. Such that, for biasing $\mu_L = eV$ and $\mu_R = 0$, all three processes, the ET, LAR, and CAR, contribute to the transport, leading thus to an enlarged differential conductance amplitude. The CAR processes contribution to dI_L/dV is finite except the case when $\mu_L = -\mu_R = eV/2$, as we have discussed in Sect. 2.2.2.

Particularly, we investigate the influence of an asymmetry in QD-lead couplings on differential conductance periodicity. To do this, we consider $\Gamma_L \neq \Gamma_R$, with electron-hole symmetry in the QD-electrode couplings $\Gamma_\alpha^e = \Gamma_\alpha^h = \Gamma_\alpha$, and arbitrary biases $\mu_L = qeV, eV = \mu_L - \mu_R$, similarly to Fig. 2.6. In Fig. 2.7a–b the differential conductance for left lead dI_L/dV and in Fig. 2.7c–d the CAR component of the differential conductance for left lead dI_L^{CAR}/dV are plotted against the magnetic flux phase ϕ, where Γ_R is a constant and Γ_L is varied at zero temperature. We find that the dot-lead couplings, Γ_L and Γ_R, mostly modify the magnitude and shape of the differential conductance, but have no impact on its periodicity. We also observe that the magnitude of

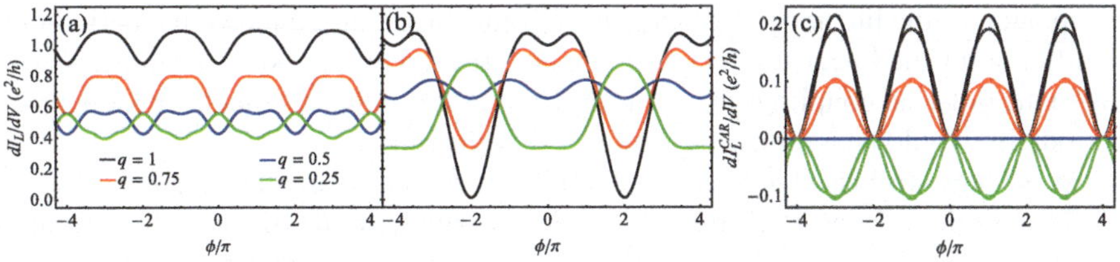

Fig. 2.6 The zero-temperature differential conductance for left (L) lead dI_L/dV versus magnetic flux phase ϕ for **a** nonoverlapping ($\varepsilon_M = 0\,\Gamma$) and **b** overlapping MZMs with energy $\varepsilon_M = 0.3\,\Gamma$. **c** The CAR component of the zero-temperature differential conductance for left lead dI_L^{CAR}/dV, where the solid lines represent the results for $\varepsilon_M = 0.3\,\Gamma$, while the dashed lines refer to the results for $\varepsilon_M = 0\,\Gamma$ case, respectively. The QD is biased for different $\mu_L = qeV$ values with $eV = 0.28\,\Gamma$ such that $\mu_L - \mu_R = eV$. The rest of parameters are chosen as: the dot-MZM couplings $|\lambda_1| = |\lambda_2| = 0.3\,\Gamma$ and QD energy $\varepsilon_d = 0\,\Gamma$. Reprinted with permission from [159]. Copyright 2022, American Physical Society

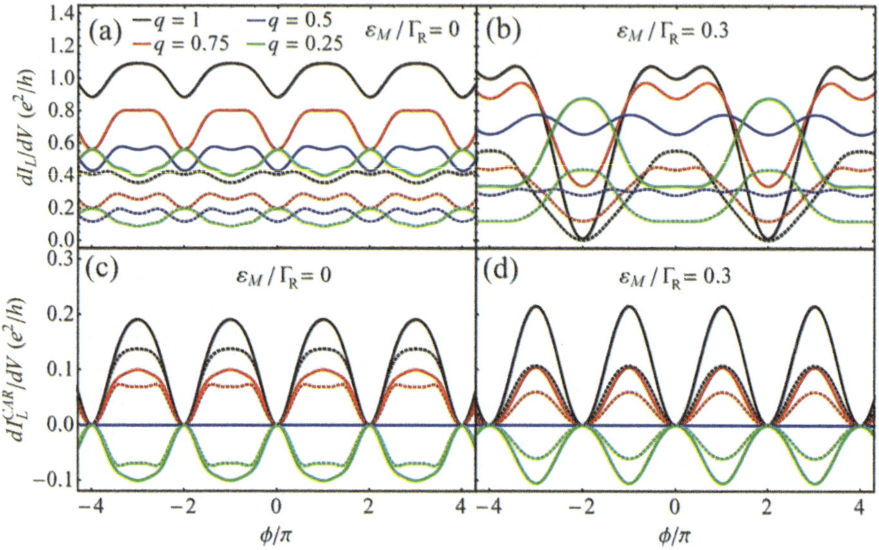

Fig. 2.7 The zero-temperature differential conductance dI_L/dV for left (L) lead against the magnetic flux phase ϕ for an asymmetrically coupled QD-leads system ($\Gamma_L \neq \Gamma_R$) with: **a** unhybridized ($\varepsilon_M = 0\,\Gamma_R$) and **b** hybridized Majoranas of overlap energy $\varepsilon_M = 0.3\,\Gamma_R$. The CAR component of the zero-temperature differential conductance dI_L^{CAR}/dV for lead L with: **c** $\varepsilon_M = 0\,\Gamma_R$ and **d** $\varepsilon_M = 0.3\,\Gamma_R$, respectively. The QD is biased for different $\mu_L = qeV$ values with $eV = 0.28\,\Gamma_R$ such that $eV = \mu_L - \mu_R$. The solid lines correspond to $\Gamma_L = \Gamma_R$, while the dashed lines to $\Gamma_L = 0.2\,\Gamma_R$, respectively. The rest of parameters are chosen as: the symmetrical dot-MZM couplings $|\lambda_1| = |\lambda_2| = 0.3\,\Gamma_R$ and dot energy $\varepsilon_d = 0\,\Gamma_R$

CAR component of the differential conductance decreases with decreasing dot-lead coupling Γ_L and again vanishes for the choice of the potential $\mu_L = -\mu_R$ ($q = 1/2$). The linear conductance reads $\mathcal{G}_L = \frac{e^2}{h}[\mathcal{T}_{LR}^{ee}(0) + 2q\mathcal{T}_{LL}^{eh}(0) + (2q-1)\mathcal{T}_{LR}^{eh}(0)]$. We see that, in the case of $q = 1/2$, on which we mainly focus in this chapter, the CAR component of the linear conductance indeed vanishes. Consequently, the periodicity of differential and linear conductances are not affected by a possible asymmetry arising in the dot-lead couplings ($\Gamma_L \neq \Gamma_R$). This further indicates that the conductance periodicity would be a robust signature in experiments.

As we have discussed in Sect. 2.2.2, for a choice of system biasing $\mu_L = -\mu_R = eV/2$, only the ET and LAR processes will contribute to the linear and differential conductances. Therefore, we now explore the ET and LAR components of the zero-temperature linear conductance $\mathcal{G}$ when the coupling $|\lambda_2|$ is varied with fixed $|\lambda_1|$. The results are shown in Fig. 2.8a for the total zero-temperature linear conductance $\mathcal{G}$ with its ET ($\mathcal{G}_{ET}$) and LAR ($\mathcal{G}_{LAR}$) components versus magnetic flux phase ϕ for degenerate MBSs ($\varepsilon_M = 0$) when the QD energy is $\varepsilon_d = 0$. We observe that the ET ($\mathcal{G}_{ET}$) and LAR ($\mathcal{G}_{LAR}$) linear conductances display the same properties to $\mathcal{G}$ discussed in Fig. 2.5a–c. That is, $\mathcal{G}_{ET}$ and $\mathcal{G}_{LAR}$ oscillate with a period of 2π, showing maxima at magnetic flux phases $\phi = (2n+1)\pi$, with amplitudes $\mathcal{G}_{ET} =$

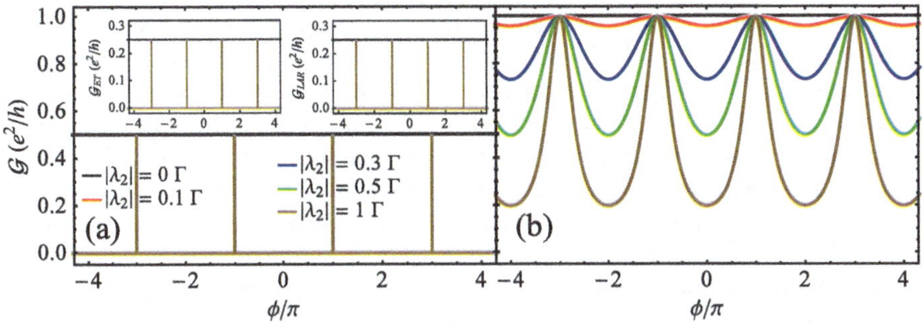

Fig. 2.8 **a** The zero-temperature linear conductance $\mathcal{G}$ versus magnetic flux phase ϕ for unhybridized MBSs ($\varepsilon_M = 0\,\Gamma$). The left and right insets refer to the results for the ET ($\mathcal{G}_{ET}$) and LAR ($\mathcal{G}_{LAR}$) components of $\mathcal{G}$, as a function of magnetic flux phase ϕ. **b** The zero-temperature linear conductance $\mathcal{G}$ versus magnetic flux phase ϕ for hybridized MZMs with overlap energy $\varepsilon_M = 0.3\,\Gamma$. The dot-MZM coupling strength $|\lambda_1|$ is fixed as $|\lambda_1| = 0.3\,\Gamma$, and the dot-MZM coupling strength $|\lambda_2|$ is varied. The dot energy is $\varepsilon_d = 0\,\Gamma$. Reproduced with permission from [159]. Copyright 2022, American Physical Society

$\mathcal{G}_{LAR} = e^2/4h$, independent of the finite values of $|\lambda_2|$. The total linear conductance $\mathcal{G}$ reaches the $e^2/2h$ value. When the dot connects to one MZM (i.e., $|\lambda_2| = 0$), the ET and LAR components of the linear conductance take the same $e^2/4h$ value independent of the magnetic flux phase ϕ. At $\phi \neq (2n+1)\pi$, the linear conductances exhibit zero as long as the dot-MZM couplings $|\lambda_j|$ are finite. Note that the zero-temperature linear conductances are independent of the values of dot energy ε_d. Due to the fact that $\mathcal{G}_{ET} \propto |G^r_{d11}(0)|^2$ and $\mathcal{G}_{LAR} \propto |G^r_{d12}(0)|^2$ at zero temperature, the behavior of ET and LAR linear conductances can be analytically verified by using the relations (2.29a)–(2.29c). If $\varepsilon_M = 0$ and $|\lambda_j| \neq 0$, the retarded Green's functions in the limit $\varepsilon \to 0$ read

$$G^r_{d11}(\varepsilon \to 0) \approx \left[2(\varepsilon + i\Gamma) - \frac{4}{\varepsilon}\frac{|\lambda_1\lambda_2|^2\cos^2\frac{\phi}{2}}{|\lambda_1|^2 + |\lambda_2|^2}\right]^{-1},$$
$$G^r_{d12}(\varepsilon \to 0) \approx \frac{\frac{|\lambda_1|^2 - |\lambda_2|^2}{|\lambda_1|^2 + |\lambda_2|^2}\cos\frac{\phi}{2} - i\sin\frac{\phi}{2}}{2(\varepsilon + i\Gamma) - \frac{4}{\varepsilon}\frac{|\lambda_1\lambda_2|^2\cos^2\frac{\phi}{2}}{|\lambda_1|^2 + |\lambda_2|^2}}. \tag{2.33}$$

We observe from Eq. (2.33) that the Green's functions are equal to $G^r_{d11}(\varepsilon \to 0) = G^r_{d12}(\varepsilon \to 0) = 0$ at $\phi \neq (2n+1)\pi$, but at $\phi = (2n+1)\pi$, they take the form $G^r_{d11}(\varepsilon \to 0) \approx 1/2(\varepsilon + i\Gamma)$ and $G^r_{d12}(\varepsilon \to 0) \approx i(-1)^{n+1}/2(\varepsilon + i\Gamma)$, regardless the QD energy and finite values of dot-MZM couplings $|\lambda_j|$. For $|\lambda_1| \neq 0$ and $|\lambda_2| = 0$, the retarded Green's functions in Eq. (2.33) become $G^r_{d11}(\varepsilon \to 0) \approx 1/2(\varepsilon + i\Gamma)$ and $G^r_{d12}(\varepsilon \to 0) \approx e^{-i\phi/2}/2(\varepsilon + i\Gamma)$, regardless the value of ε_d, which leads to $\mathcal{G}_{ET} = \mathcal{G}_{LAR} = e^2/4h$. Note that the ε_d-independence of zero-temperature linear conductance for $\varepsilon_M = 0$ was also reported in the literature for different devices based on QDs coupled to TSNWs [2, 5, 36]. When the Majorana overlap energy $\varepsilon_M \neq 0$

and the QD level $\varepsilon_d = 0$ at zero temperature [see Fig. 2.8b], the peaks of conductance occur at $\phi = (2n+1)\pi$. For $\varepsilon_M \neq 0$, the retarded Green's function becomes $G^r_{d12}(\varepsilon \to 0) \approx 0$, regardless the values of QD energy ε_d, which thus results in the disappearance of LAR conductance, $\mathcal{G}_{LAR} \approx 0$. Consequently, the zero-temperature linear conductance $\mathcal{G}$ is composed from the ET conductance, i.e., $\mathcal{G} \approx \mathcal{G}_{ET}$, and it presents a 2π periodicity in magnetic flux phase at $\varepsilon_d = 0$. We see that with the enhancement of coupling $|\lambda_2|$, the minima of linear conductance $\mathcal{G}$ emerging at the $\phi = 2n\pi$ values gradually reduce and the resonances start to narrow. Contrary to the nonoverlapping Majoranas case, the maxima of $\mathcal{G}$ for hybridized MZMs reach now the value e^2/h. As a consequence, for degenerate Majoranas, the zero-temperature linear conductance is composed in equal manner from the nonlocal (ET) and local (LAR) processes, regardless the values of ε_d. When the MZMs overlap, the zero-temperature linear conductance $\mathcal{G}$ is only attributed to the nonlocal (ET) processes such that the contribution of local (LAR) processes is vanished.

To further study the behavior of $\mathcal{G}$ in the regime $\varepsilon_M \neq 0$ and $\varepsilon_d \neq 0$, the zero-temperature total linear conductance, i.e., $\mathcal{G} \approx \mathcal{G}_{ET}$ is plotted in Fig. 2.9 as a function of magnetic flux phase ϕ and dot energy ε_d when the dot-MZM coupling strength $|\lambda_1|$ is fixed and the dot-MZM coupling $|\lambda_2|$ is varied. We see that the linear conductance $\mathcal{G}$ significantly depends on magnetic flux phase ϕ and QD energy ε_d. It was found that $\mathcal{G}$ strongly depends on the QD level ε_d when the Majorana overlap energy ε_M surpasses the temperature in system, i.e., $\varepsilon_M \gtrsim T$ [36]. In the case where the QD energy is pinned at the Fermi level, i.e., $\varepsilon_d = 0$, the linear conductance has a 2π periodicity as a function of magnetic flux phase ϕ in agreement with the results in Fig. 2.8b. When the dot energy deviates from the Fermi level, i.e., $\varepsilon_d \neq 0$, then the linear conductance $\mathcal{G}$ exhibits a 4π periodicity in magnetic flux phase ϕ, independent of the finite values of QD-MBS coupling $|\lambda_2|$. The $\mathcal{G}$ map details further modifies with the grow of coupling $|\lambda_2|$.

We now investigate the magnetic flux phase ϕ and QD energy ε_d dependence of the differential conductance dI/dV at zero temperature in Fig. 2.10 when the bias voltage eV is varied and the symmetrical dot-MZM coupling strengths $|\lambda_1| = |\lambda_2|$ are fixed

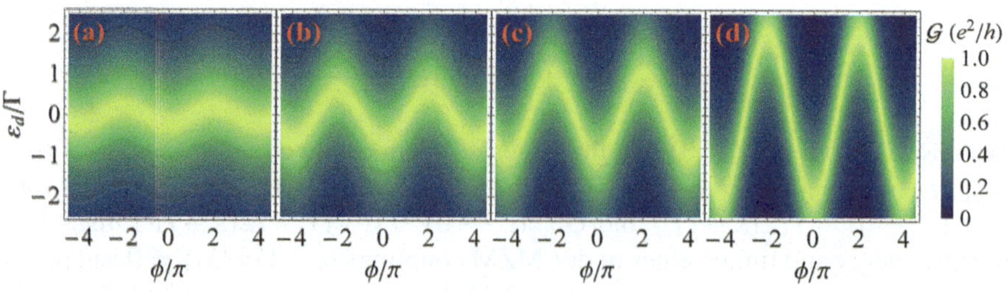

Fig. 2.9 The linear conductance $\mathcal{G}$ versus magnetic flux phase ϕ and QD energy ε_d at zero temperature for hybridized Majoranas with overlap energy $\varepsilon_M = 0.3\,\Gamma$ and fixed dot-MBS coupling $|\lambda_1| = 0.3\,\Gamma$ at different dot-MBS coupling strengths $|\lambda_2|$: **a** $|\lambda_2| = 0.1\,\Gamma$, **b** $|\lambda_2| = 0.3\,\Gamma$, **c** $|\lambda_2| = 0.5\,\Gamma$ and **d** $|\lambda_2| = 1\,\Gamma$. Reprinted with permission from [159]. Copyright 2022, American Physical Society

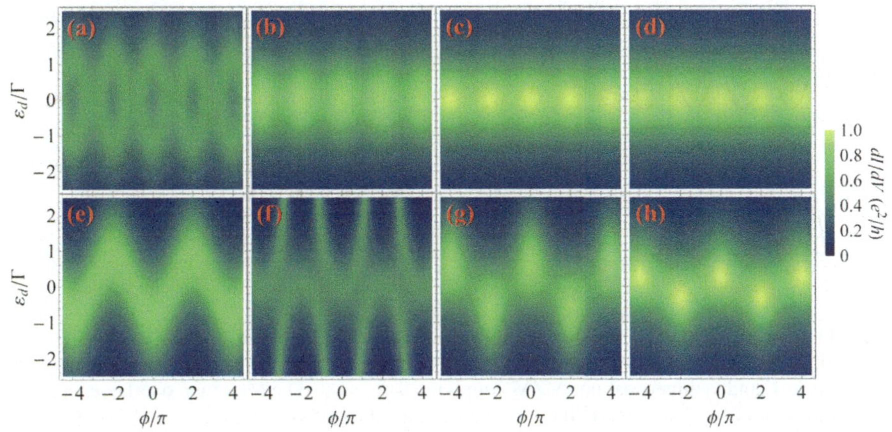

Fig. 2.10 The differential conductance dI/dV versus magnetic flux phase ϕ and QD energy ε_d at zero temperature for: **a–d** nonoverlapping ($\varepsilon_M = 0\,\Gamma$) and **e–h** overlapping MBSs with energy $\varepsilon_M = 0.3\,\Gamma$ and symmetrical dot-MZM couplings $|\lambda_1| = |\lambda_2| = 0.3\,\Gamma$. The bias voltage eV is chosen as: **a, e** $eV = 0.28\,\Gamma$; **b, f** $eV = 0.55\,\Gamma$; **c, g** $eV = 0.85\,\Gamma$; and **d, h** $eV = 1.05\,\Gamma$. Reprinted with permission from [159]. Copyright 2022, American Physical Society

for both unhybridized and hybridized MZMs. We see that the differential conductance shows a 2π periodicity in magnetic flux phase ϕ when the Majorana wavefunctions do not overlap ($\varepsilon_M = 0$), regardless the QD level ε_d tuning [Fig. 2.10a–d]. For hybridized MBSs, the 2π periodicity of dI/dV in ϕ transforms to a 4π one when the dot energy ε_d deviates from the Fermi level, i.e., $\varepsilon_d \neq 0$ [Fig. 2.10e–h]. Notice that the maps of differential conductance for hybridized MBSs are strongly sensitive to small variations of dot level ε_d and bias voltage eV. Note that the nonlocal (ET) and local (LAR) components of the differential conductance dI/dV oscillate with a 2π period in magnetic flux phase ϕ for degenerate MBSs independent of the ε_d values. In addition, the nonlocal (ET) component of the differential conductance dominates compared to the local (LAR) one.

We study the influence of finite temperature T on the spectra of linear and differential conductance. To do this, we plot, in Fig. 2.11, the linear conductance $\mathcal{G}$ against the magnetic flux phase ϕ and QD energy ε_d at different values of the temperature T. We consider the cases of unhybridized ($\varepsilon_M = 0$) and hybridized ($\varepsilon_M \neq 0$) Majoranas when the QD symmetrically connects to both MZMs with different values of the symmetrical dot-MZM couplings $|\lambda_j| = |\lambda|$. We observe that for nonoverlapping Majoranas, the linear conductance takes the $e^2/2h$ value at zero temperature and magnetic flux phase $\phi = (2n + 1)\pi$, regardless the finite values of the couplings $|\lambda|$, as was found in Fig. 2.5a–c. The temperature displays a broadening effect on conductance resonances. The minima of linear conductance peaks significantly grow with temperature T when the dot weakly hybridizes with the MBSs, and their maxima surpass the $e^2/2h$ value [see Fig. 2.11a, c]. The linear conductance increases with temperature and after reaching a critical value it begins to decrease again with

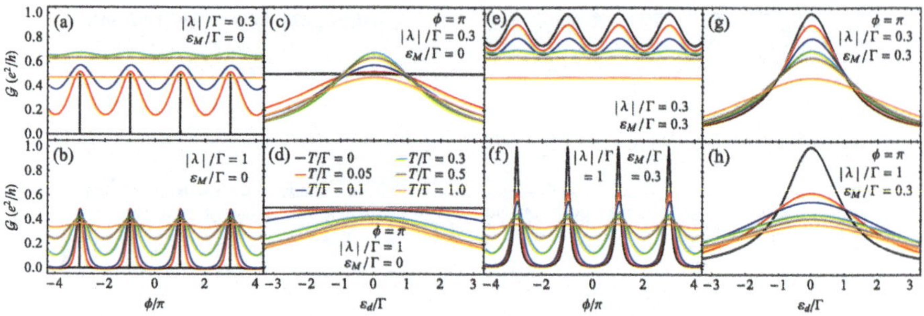

Fig. 2.11 **a, b, e, f** The linear conductance $\mathcal{G}$ versus magnetic flux phase ϕ at QD energy $\varepsilon_d = 0\,\Gamma$ and **c, d, g, h** $\mathcal{G}$ versus QD energy ε_d for magnetic flux phase $\phi = \pi$ at different values of the temperature T and symmetrical dot-MZM couplings $|\lambda_j| = |\lambda|$. The Majorana overlap energy ε_M is chosen as: **a–d** $\varepsilon_M = 0\,\Gamma$ and **e–h** $\varepsilon_M = 0.3\,\Gamma$. The values of symmetrical dot-MZM couplings $|\lambda|$ are: **a, c, e, g** $|\lambda| = 0.3\,\Gamma$ and **b, d, f, h** $|\lambda| = 1\,\Gamma$. Reprinted with permission from [159]. Copyright 2022, American Physical Society

further increase in T, as was also found in the literature [8, 40]. At stronger $|\lambda|$, the peaks of $\mathcal{G}$ are broadened at higher temperatures but are narrower than those that appear for weak dot-MZM couplings $|\lambda|$. The minima of linear conductance are less sensitive to the changes in T with stronger $|\lambda|$ coupling than with weak couplings at low temperatures. The amplification in magnitude of $\mathcal{G}$ maxima with temperature is not observable for stronger $|\lambda|$ and the conductance peaks no longer exceed the value $e^2/2h$ [Fig. 2.11b, d]. In conclusion, the linear conductance presents a non-monotonic behavior with T for weak couplings $|\lambda|$ by exceeding the value $e^2/2h$, while for stronger $|\lambda|$, $\mathcal{G}$ is limited from above by the value $e^2/2h$ and decays with T. The finite Majorana overlap energy ($\varepsilon_M \neq 0$) dramatically modifies the spectrum of $\mathcal{G}$ as is illustrated in Fig. 2.11e–h. For hybridized MBSs, the maxima of linear conductance reach the e^2/h value at zero temperature, independent of the finite values of the couplings $|\lambda|$. The peaks of $\mathcal{G}$ do not surpass the value e^2/h at any finite QD-MZM couplings or temperature considered here. The linear conductance rapidly decreases with growing temperature for strong $|\lambda|$ couplings. Contrary to the unhybridized Majoranas case, the maxima of linear conductance $\mathcal{G}$ do not grow with T for hybridized MBSs. Without showing numerical results, we note that dI/dV responds similarly to the changes in T as $\mathcal{G}$ presented in Fig. 2.11. Namely, for $\varepsilon_M = 0$, with weak dot-Majorana couplings, the local minima emerging at magnetic flux phase $\phi = (2n+1)\pi$ [Fig. 2.5d] enhance with T and change to plateaus. By further increasing the value of temperature results in the transformation of dI/dV plateaus into maxima. The magnitude of these conductance peaks increases with temperature, and after reaching a critical value, it drops with further increase in T. The increase in symmetrical dot-MZM coupling $|\lambda|$ [see Fig. 2.5e, f] leads to the local minima of differential conductance transform into sharper resonances whose amplitude reduces on increasing T. For hybridized MBSs, with both weak and strong dot-MZM coupling strengths $|\lambda|$, the amplification of differential conductance resonances with temperature is not seen, as in the case of linear conductance.

2.3.2 Results in Presence of Electron-Phonon Interaction

In this section, we analyze the influence of EPI on transport characteristics of the QD-MBS system. Such devices, for which a quantum system is attached to an environment, play a central role in transport experiments since they can induce decoherence effects on Majorana-based quantum computation [126, 133–136, 138]. In the present case, we study the phonon-assisted MBS-induced transport properties by exploring the normalized equilibrium spectral function $\Gamma\mathcal{A}_d(\varepsilon)$, tunneling current, linear and differential conductances at different temperatures. Note that, in the presence of EPI, the spectral function of the QD, $\mathcal{A}_d(\varepsilon)$, given in Eq. (2.13), depends on the bias voltage eV and temperature T through the Fermi-Dirac distribution functions, emerging in the Keldysh equations (2.28). We recall that in the presence of EPI the QD energy, the dot-lead and dot-MZM couplings are renormalized: $\tilde{\varepsilon}_d = \varepsilon_d - \beta^2/\omega_0$, $\tilde{\Gamma} = \Gamma e^{-g(2N_{\rm ph}+1)}$ and $|\tilde{\lambda}_j| = |\lambda_j|e^{-g(N_{\rm ph}+1/2)}$, respectively.

We first investigate in Fig. 2.12 the normalized equilibrium ($eV = 0$) spectral function of the dot $\Gamma\mathcal{A}_d(\varepsilon)$ when the QD symmetrically couples to both the ends of a TSNW with different coupling strengths $|\lambda_j| = |\lambda|$ and three magnetic flux phase ϕ values for unhybridized Majoranas with a fixed EPI strength $\beta = 2.5\,\Gamma$, at dot energy $\varepsilon_d = 0$ and zero temperature. The normalized equilibrium spectral function $\Gamma\mathcal{A}_d(\varepsilon)$ without EPI and MZMs (represented by the dot-dashed black line) manifests itself as a resonant peak of Lorentzian lineshape at energy $\varepsilon = \varepsilon_d$ [162]. The presence of EPI ($\beta \neq 0$) induces polaronic effects and leads to renormalized parameters related to the QD. We observe that in the absence of Majoranas ($|\lambda| = 0$, denoted by the solid black line in Fig. 2.12), the resonant peak gets redshifted by showing up at the renormalized dot energy $\varepsilon = \tilde{\varepsilon}_d = \varepsilon_d - g\omega_0$ with the dimensionless coupling $g = (\beta/\omega_0)^2$. In addition, beside the main resonance, a series of satellite peaks emerge in the spectrum of $\Gamma\mathcal{A}_d(\varepsilon)$ at energies $\varepsilon = \tilde{\varepsilon}_d \pm p\omega_0$ (where p is an integer) by opening additional phonon-assisted transport channels [69, 142, 147, 162]. These phonon-assisted side-bands in the $\Gamma\mathcal{A}_d(\varepsilon)$ spectrum correspond to electron tunnelings through the dot by emission or absorption of phonons at frequency ω_0 [161]. At zero temperature, the Franck-Condon factor $\mathcal{L}_p$ vanishes for $p < 0$, thus the side-peaks emerging in the negative energy domain result from $\tilde{G}_d^{<}(\varepsilon + p\omega_0)$ in relation (2.27), while the peaks from the positive energy domain come from $\tilde{G}_d^{>}(\varepsilon - p\omega_0)$ [162]. The approximation of Fermi-Dirac distribution function at zero temperature with the Heaviside step function results in a discontinuity of the shape of side-peaks. Note that the side-peaks height reduces at large p and vanishes completely away from the main resonance. The phonon side-bands developed in the vicinity of central resonance are not symmetric with respect to $\tilde{\varepsilon}_d$ [162]. As we have seen in Fig. 2.2, at magnetic flux phase $\phi = 0$ without EPI when the QD connects to both MBSs of the TSNW, two maxima emerge in the spectrum at energies $\varepsilon \approx \pm\sqrt{2}\lambda$. In addition, the zero-energy normalized spectral function equals zero, $\Gamma\mathcal{A}_d(0) = 0$, regardless the finite values of the symmetrical coupling $|\lambda|$ with $\beta = 0$. In the $\beta \neq 0$ case, the $\Gamma\mathcal{A}_d(\varepsilon)$ spectrum is clearly modified [see Fig. 2.12a]. We observe that the main resonance and the phonon-assisted satellite peaks located in the negative energy region of $\Gamma\mathcal{A}_d(\varepsilon)$ are

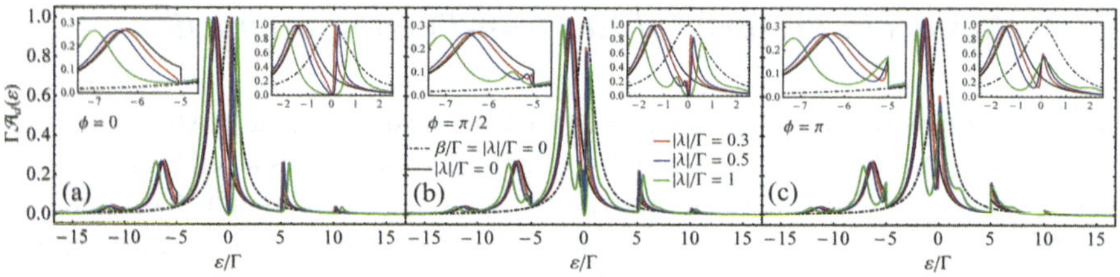

Fig. 2.12 The QD normalized equilibrium spectral function $\Gamma\mathcal{A}_d(\varepsilon)$ at zero temperature with EPI of strength $\beta = 2.5\,\Gamma$ when the dot is attached to both MZMs with different symmetrical dot-MBS couplings $|\lambda_j| = |\lambda|$ for nonoverlapping MBSs ($\varepsilon_M = 0\,\Gamma$), at dot energy $\varepsilon_d = 0\,\Gamma$. The magnetic flux phase is chosen as: **a** $\phi = 0$, **b** $\phi = \pi/2$, and **c** $\phi = \pi$. Here, the right inset in each panel zoom in on the zero-energy features in the normalized equilibrium spectral function. The left inset shows a zoom at the first phonon-induced satellite peak in negative energy domain. Reproduced with permission from [159]. Copyright 2022, American Physical Society

redshifted, while those that develop in the positive energy domain are blueshifted on increasing $|\lambda|$. Note that the zero-energy normalized spectral function $\Gamma\mathcal{A}_d(0)$ remains unaffected, i.e., $\Gamma\mathcal{A}_d(0) = 0$. In the presence of magnetic flux and EPI [see Fig. 2.12b, c], the QD normalized spectral function becomes more complicated. Remarkably, the value of zero-energy normalized spectral function $\Gamma\mathcal{A}_d(0)$ remains unchanged under EPI, i.e., and it takes 0 at $\phi = \pi/2$ and $1/2$ at $\phi = \pi$, respectively. At magnetic flux phase $\phi = \pi/2$, the four-peak structure of $\Gamma\mathcal{A}_d(\varepsilon)$, which appears for strong symmetrical QD-MBS couplings $|\lambda|$ in the $\beta = 0$ limit (see Fig. 2.2), is highly deformed in the $\beta \neq 0$ case. We can see that at $\phi = \pi$, the three-peak structure in $\Gamma\mathcal{A}_d(\varepsilon)$ near $\varepsilon \approx 0$, associated to the Majorana signature, shows up only for strong symmetrical dot-MZM couplings $|\lambda|$.

We study in Fig. 2.13a, b the effect of Majorana overlap energy ε_M on the normalized equilibrium spectral function of the dot $\Gamma\mathcal{A}_d(\varepsilon)$ at two magnetic flux phases ϕ when the QD is connected to both the ends of a TSNW with weak symmetrical couplings $|\lambda_1| = |\lambda_2|$, at QD energy level $\varepsilon_d = 0$ and zero temperature in the presence of EPI. In the absence of EPI ($\beta = 0$) and magnetic flux ($\phi = 0$), the normalized equilibrium spectral function shows an antiresonance point at $\varepsilon \approx -\varepsilon_M$, in agreement with the results in Fig. 2.3. In the $\beta \neq 0$ case [see Fig. 2.13a for $\phi = 0$], we can observe that the antiresonance point location is immune to the presence of EPI and the two main resonances located in the vicinity of the antiresonance point are redshifted. The position of the broadened resonance is determined by the EPI coupling strength via $\tilde{\varepsilon}_d = \varepsilon_d - \beta^2/\omega_0$ and is weakly influenced by the Majorana overlap energy ε_M. The narrow peak location strongly depends on the overlap energy. The phonon-assisted satellite peaks are also affected by the finite Majorana overlap energy. In the positive energy domain ($\varepsilon > 0$), the side-peaks' height reduces with the enhancement of ε_M. In the negative energy region ($\varepsilon < 0$), the satellite resonances split into two peaks, a narrow and a broadened one. When the magnetic flux phase is equal to $\phi = \pi$ [see Fig. 2.13b], the two-peak structure located in the vicinity of the

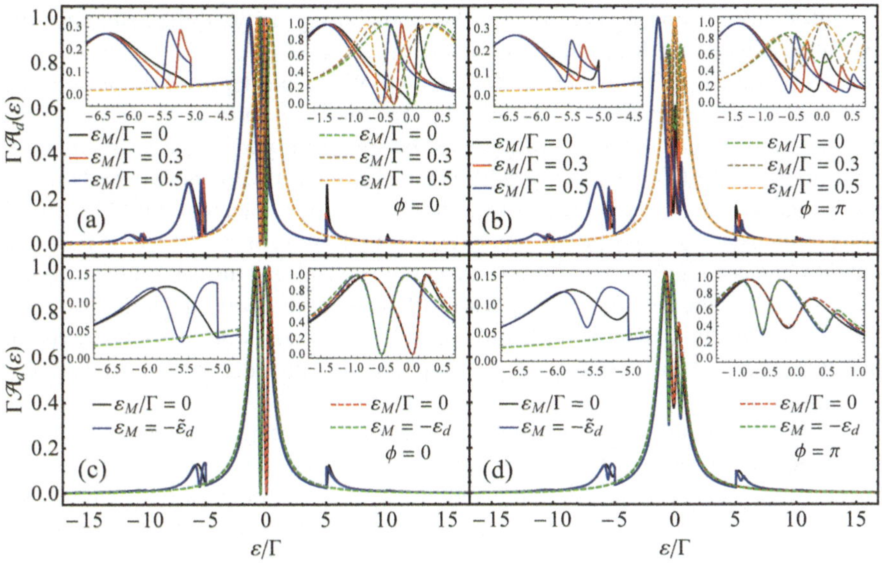

Fig. 2.13 **a, b** The normalized equilibrium spectral function of the dot $\Gamma \mathcal{A}_d(\varepsilon)$ with EPI of coupling strength $\beta = 2.5\,\Gamma$ at different Majorana overlap energies ε_M and zero temperature when the dot connects to both MZMs with $|\lambda_1| = |\lambda_2| = 0.3\,\Gamma$ and $\varepsilon_d = 0\,\Gamma$. The magnetic flux phase is: **a** $\phi = 0$ and **b** $\phi = \pi$. The solid and dashed lines correspond to the results with and without EPI. **c, d** The dot normalized equilibrium spectral function $\Gamma \mathcal{A}_d(\varepsilon)$ at zero temperature for $\varepsilon_M = 0\,\Gamma$ and $\varepsilon_M = 0.5\,\Gamma$ when the dot connects to both MZMs at $|\lambda_1| = |\lambda_2| = 0.3\,\Gamma$. In the presence of EPI (see solid lines), at QD level $\varepsilon_d = 0\,\Gamma$, the electron-phonon coupling strength is calculated as $\beta = \sqrt{\varepsilon_M \omega_0}$ with overlap energy $\varepsilon_M = 0.5\,\Gamma$ giving $\beta \approx 1.58\,\Gamma$. The dashed lines denote the results without EPI at $\varepsilon_d = -0.5\,\Gamma$. The magnetic flux phase ϕ is chosen as: **c** $\phi = 0$ and **d** $\phi = \pi$. Here, the right inset in each panel zoom in on the zero-energy features in $\Gamma \mathcal{A}_d(\varepsilon)$. The left inset shows a zoom at the first phonon-induced satellite peak in negative energy domain. Reproduced with permission from [159]. Copyright 2022, American Physical Society

zero-energy in $\Gamma \mathcal{A}_d(\varepsilon)$ transforms into a three-peak one for finite Majorana overlap energies, providing thus the fingerprints of Majoranas. In the $\varepsilon_M = 0$ case, the zero-energy normalized spectral function of the QD is immune to the presence of EPI, i.e., $\Gamma \mathcal{A}_d(\varepsilon = 0) = 1/2$. For hybridized Majoranas, the zero-energy normalized spectral function now differs from the value 1, i.e., $\Gamma \mathcal{A}_d(\varepsilon = 0) \neq 1$. Furthermore, the main narrow peak associated to the Majoranas is moved to lower energies, and its height is also decreased relative to the case without EPI. The broadened resonance again weakly depends on the overlap energy ε_M and reaches the unitary limit. The location of antiresonance points, that originally emerged without EPI at energies $\varepsilon \approx \pm\varepsilon_M$, is modified at finite β. We observe that the three-peak structures occurring at side-resonances are truncated at energies $\varepsilon \approx \pm|p|\omega_0$, and also local minima develop in $\Gamma \mathcal{A}_d(\varepsilon)$ at $\varepsilon \approx \pm(|p|\omega_0 + \varepsilon_M)$. We now focus on the $\tilde{\varepsilon}_d = -\varepsilon_M$ case with $\varepsilon_d = 0$ when the following relation can be written down $\beta = \sqrt{\varepsilon_M \omega_0}$. For the overlap energy $\varepsilon_M = 0.5\,\Gamma$ considered here, we obtain $\beta \approx 1.58\,\Gamma$, which results in a sufficiently strong β (i.e., $\beta > \Gamma$). We show in Fig. 2.13c, d the results for this special case when the Majorana overlap energy is $\varepsilon_M = -\tilde{\varepsilon}_d$ with dot level $\varepsilon_d = 0$ for $\beta \neq 0$ and with QD energy $\varepsilon_d \neq 0$ for $\beta = 0$. When the magnetic flux is switched off [$\phi = 0$, Fig. 2.13c] with $\varepsilon_M = -\tilde{\varepsilon}_d$, the two main resonances around the zero-energy $\varepsilon \approx 0$, associated to the Majoranas, become nearly identical, as seen before in the absence of EPI (see also Fig. 2.3). When the magnetic flux is switched on with $\phi = \pi$ and $\varepsilon_M = -\tilde{\varepsilon}_d$ [see Fig. 2.13d], the main sharp resonance of the three-peak structure presents an appropriate match relative to the case of $\beta = 0$, as for $\phi = 0$. In addition, the symmetry of the two resonances located around the central resonance is broken even for the case $\tilde{\varepsilon}_d = -\varepsilon_M$ with EPI, as was before discussed in Fig. 2.3 in the $\beta = 0$ limit. Note that the truncation of phonon-assisted side-peaks also exists in this special case. As a consequence, the Majorana signature depends on the variation of β, ε_M or ε_d.

Figure 2.14 illustrates the normalized equilibrium spectral function $\Gamma\mathcal{A}_d(\varepsilon)$ at zero temperature in the $\beta \neq 0$ case when varying the renormalized QD energy $\tilde{\varepsilon}_d = \varepsilon_d - \beta^2/\omega_0$ for nonoverlapping and overlapping Majoranas, with symmetrical dot-MZM coupling strengths $|\lambda_1| = |\lambda_2|$, for different values of magnetic flux phases

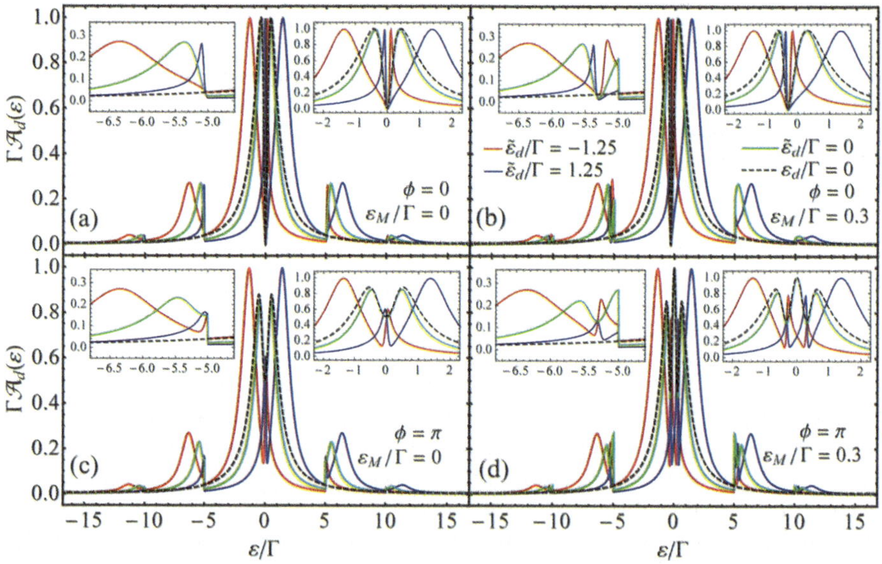

Fig. 2.14 The normalized equilibrium spectral function of the dot $\Gamma\mathcal{A}_d(\varepsilon)$ at zero temperature in the presence of EPI with electron-phonon coupling strength $\beta = 2.5\,\Gamma$ (solid lines) at different renormalized QD energies $\tilde{\varepsilon}_d$ with nonoverlapping and overlapping MBSs. The QD symmetrically couples to both Majoranas with coupling $|\lambda_1| = |\lambda_2| = 0.3\,\Gamma$ at different values of the magnetic flux phase ϕ. The dashed line shows the results for the $\beta = 0\,\Gamma$ case. The rest of the parameters are: **a** $\phi = 0$, $\varepsilon_M = 0\,\Gamma$; **b** $\phi = 0$, $\varepsilon_M = 0.3\,\Gamma$; **c** $\phi = \pi$, $\varepsilon_M = 0\,\Gamma$; and **d** $\phi = \pi$, $\varepsilon_M = 0.3\,\Gamma$. Here, the right inset in each panel zoom in on the zero-energy features in $\Gamma\mathcal{A}_d(\varepsilon)$. The left inset shows a zoom at the first phonon-induced satellite peak in negative energy domain. Reproduced with permission from [159]. Copyright 2022, American Physical Society

ϕ. The characteristics of central resonances ($p = 0$) in the normalized equilibrium spectral function reproduce the findings discussed in the $\beta = 0$ limit (see Fig. 2.3), under a mapping of the QD energy $\varepsilon_d \to \tilde{\varepsilon}_d$. The antiresonance point of $\Gamma\mathcal{A}_d(\varepsilon)$ holds for flux phase $\phi = 0$ at $\varepsilon \approx -\varepsilon_M$, without or with EPI [see Fig. 2.14a, b]. At magnetic flux phase $\phi = \pi$ with Majorana overlap energy $\varepsilon_M = 0$ [see Fig. 2.14c], the zero-energy normalized spectral function reads $\Gamma\mathcal{A}_d(\varepsilon = 0) = 1/2$, and the Fano lineshape spectral asymmetry near the zero-energy $\varepsilon \approx 0$ disappears again when the QD energy equals the EPI-induced energy shift $g\omega_0$, i.e., $\varepsilon_d = \beta^2/\omega_0$ resulting in $\tilde{\varepsilon}_d = 0$. At finite values of ε_M, the three-peak structure characterizing the MBSs manifests itself [see Fig. 2.14d]. Note that the shapes of side-resonances ($p \neq 0$), are identical but mirrored at $\tilde{\varepsilon}_d \approx 0$ for unhybridized MBSs at any ϕ, or for hybridized MBSs at $\phi = (2n + 1)\pi$. The broadened resonances in $\Gamma\mathcal{A}_d(\varepsilon)$ correspond to transmission through the renormalized dot level $\tilde{\varepsilon}_d$, while the sharp peaks are associated to regular fermionic states coming from the MZMs. Note that if both ε_M and $\tilde{\varepsilon}_d$ disappear, the main resonances will be approximately symmetric at any values of ϕ [see Fig. 2.14a, c], as was mentioned above with EPI in Fig. 2.13, or without EPI in Fig. 2.3.

The influence of temperature T on the normalized equilibrium spectral function $\Gamma\mathcal{A}_d(\varepsilon)$ with EPI is analyzed in Fig. 2.15. The results are performed at different values of the temperature for nonoverlapping and overlapping Majoranas, when both the ends of the TSNW weakly connect to the QD with symmetrical couplings $|\lambda_1| = |\lambda_2|$, at dot level $\varepsilon_d = 0$, at two values of ϕ. One observes that the main resonances in $\Gamma\mathcal{A}_d(\varepsilon)$ are less sensitive to the changes in T than the phonon-assisted satellite peaks. Due to the fact that the normalized spectral function $\Gamma\mathcal{A}_d(\varepsilon)$ in Eq. (2.13) is proportional to the Franck-Condon factor $\mathcal{L}_p$, thus the height of peaks is directly affected by the coefficients $\mathcal{L}_p$. Namely, the weak dependence of $\mathcal{L}_p$ on T in the low-temperature regime considered here leads to very small changes in the peaks' height. However, at high temperatures, all the coefficients $\mathcal{L}_p$ begin to decay as $1/\sqrt{T}$ leading thus to the reduction in height of peaks. The Fermi-Dirac distribution function at finite temperatures results in a smoothing of the discontinuity, which is present at zero temperature, in the side-peaks shape. The width of peaks is affected by the renormalized dot-lead couplings $\tilde{\Gamma} = \Gamma e^{-g(2N_{\text{ph}}+1)}$ where $\tilde{\Gamma}$ decreases with the increase of temperature. This results in a narrowing of resonances with the grow of T [154]. The position of peaks is weakly affected by T via the renormalized dot-MZM couplings, $|\tilde{\lambda}| = |\lambda|e^{-g(N_{\text{ph}}+1/2)}$.

We now evaluate in Fig. 2.16 the normalized equilibrium spectral function $\Gamma\mathcal{A}_d(\varepsilon)$ at a finite temperature for different strengths of the electron-phonon coupling β, when the dot connects to both MZMs with couplings $|\lambda_1| = |\lambda_2|$, at different magnetic flux phases ϕ, for degenerate Majoranas, at QD energy level $\varepsilon_d = 0$. We first observe that the normalized equilibrium spectral function at zero-energy takes the value $\Gamma\mathcal{A}_d(0) = 0$ or $\Gamma\mathcal{A}_d(0) = 1/2$, with $\phi = 0$ or $\phi = \pi$, which is consistent with the results from Fig. 2.2. In addition, we see that for magnetic flux phase $\phi = 0$ the location of antiresonance point at energy $\varepsilon = 0$ is not influenced by the EPI strength β. Also, the main resonance emerging in the negative energy domain of $\Gamma\mathcal{A}_d(\varepsilon)$

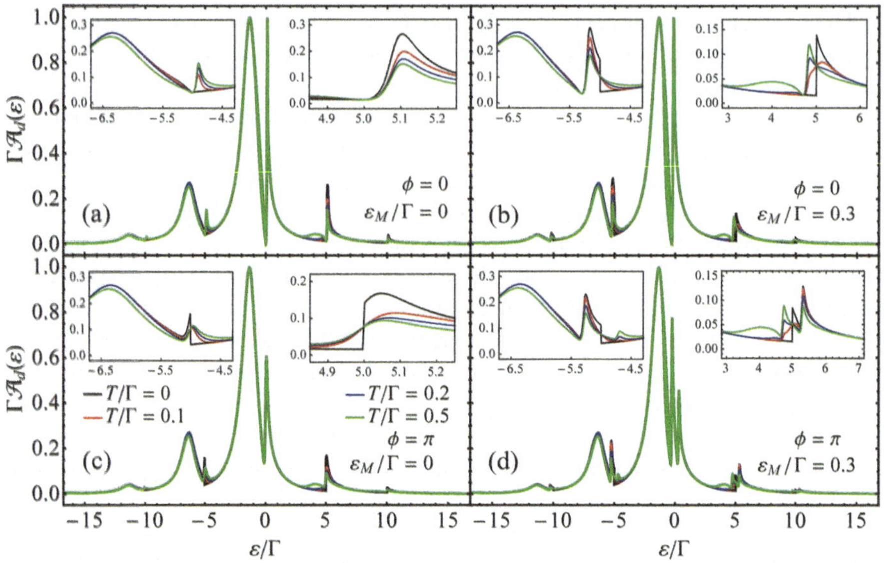

Fig. 2.15 The QD normalized equilibrium spectral function $\Gamma \mathcal{A}_d(\varepsilon)$ with EPI at different values of the temperature T and magnetic flux phase ϕ for unhybridized and hybridized Majoranas when the dot connects to both MZMs with $|\lambda_1| = |\lambda_2| = 0.3\,\Gamma$. The EPI strength and QD energy are $\beta = 2.5\,\Gamma$ and $\varepsilon_d = 0\,\Gamma$, respectively. The parameters are chosen as: **a** $\phi = 0$, $\varepsilon_M = 0\,\Gamma$; **b** $\phi = 0$, $\varepsilon_M = 0.3\,\Gamma$; **c** $\phi = \pi$, $\varepsilon_M = 0\,\Gamma$; and **d** $\phi = \pi$, $\varepsilon_M = 0.3\,\Gamma$. The left and right insets in each panel zoom in on the first phonon-assisted peak at negative and positive energy, respectively. Reproduced with permission from [159]. Copyright 2022, American Physical Society

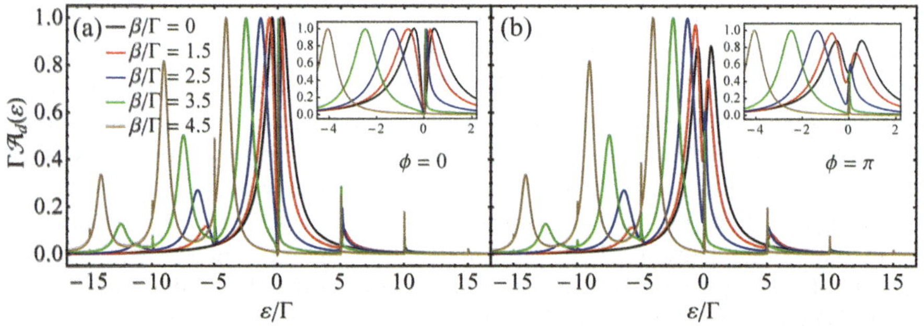

Fig. 2.16 The QD normalized equilibrium spectral function $\Gamma \mathcal{A}_d(\varepsilon)$ in the presence of EPI with different electron-phonon coupling strength β values for nonoverlapping Majoranas ($\varepsilon_M = 0\,\Gamma$), when the QD couples to both MBSs with symmetrical couplings $|\lambda_1| = |\lambda_2| = 0.3\,\Gamma$ at QD energy $\varepsilon_d = 0\,\Gamma$ and a finite temperature $T = 0.1\,\Gamma$. The magnetic flux phase is set as: **a** $\phi = 0$ and **b** $\phi = \pi$. The inset in each panel is a zoom in on the zero-energy features in $\Gamma \mathcal{A}_d(\varepsilon)$. Reproduced with permission from [159]. Copyright 2022, American Physical Society

moves to the lower energy values with the increase in β because of the renormalized dot energy, i.e., $\tilde{\varepsilon}_d = \varepsilon_d - \beta^2/\omega_0$. The central resonance showing up in the positive energy region of $\Gamma\mathcal{A}_d(\varepsilon)$, caused by Majoranas, shifts towards the zero-energy with the enhancement of EPI strength β. The location of satellite peaks in $\Gamma\mathcal{A}_d(\varepsilon)$ is affected by β. Taking into account that the dot-electrode couplings Γ are renormalized when the QD couples to the phonon mode, i.e., $\tilde{\Gamma} = \Gamma e^{-g(2N_{\rm ph}+1)}$, thus at finite temperatures also the width and height of peaks are influenced by the EPI strength β as well as the temperature T. An increase in EPI strength β leads to the suppression of QD-lead couplings $\tilde{\Gamma}$, resulting thus in sharp resonances with amplified height. This causes a redistribution in the weight of side-peaks [154]. Notice that the amplification of the height of side-peaks when β gets stronger is more observable in the negative energy domain than in the positive one. In addition, $\mathcal{L}_p$ increases with the EPI strength β for each p up to a maximum value, after which it begins to decrease. Note that due to the renormalized QD-MBS couplings in presence of EPI, $|\tilde{\lambda}_j|$ decrease with the increase of β which also affects the location of peaks.

We now analyze the properties of current through a QD symmetrically coupled to normal electrodes and one of the ends of a TSNW when the localized dot electron is in interaction with a single long-wave optical phonon mode. The Hamiltonian of such system is obtained by setting $\lambda_1 = |\lambda|$ and $\lambda_2 = 0$ in Eq. (2.1). The corresponding dot retarded Green's functions are given in Eq. (2.30). We study the effect of system parameters on the current–dot energy, current–bias voltage, current–QD-MZM coupling and current–temperature characteristics in both the absence and presence of EPI.

We first investigate the influence of dot-Majorana coupling $|\lambda|$ and bias voltage eV on the $I - \tilde{\varepsilon}_d$ characteristics (where $\tilde{\varepsilon}_d = \varepsilon_d - \beta^2/\omega_0$) for degenerate Majoranas under EPI at a single finite temperature $T = 0.1\,\Gamma$. We plot, in Fig. 2.17a, the current I versus $\tilde{\varepsilon}_d$ when the dot is attached to one MBS with different values of coupling $|\lambda_1| = |\lambda|$ ($|\lambda_2| = 0$ and $\phi = 0$) at bias voltage $eV = 2\,\Gamma$ and $\varepsilon_M = 0$ with EPI of coupling strength $\beta = 2.5\,\Gamma$ and without EPI. We can see that without MBSs and EPI, a single Lorentzian peak shows up at dot energy $\tilde{\varepsilon}_d = \varepsilon_d = 0$. The magnitude of current peak is decreased when the QD connects to the MZM ($|\lambda| \neq 0$) in the QD energy region $-(\Gamma + \frac{|eV|}{2}) \lesssim \varepsilon_d \lesssim (\Gamma + \frac{|eV|}{2})$. When the QD energy is $|\varepsilon_d| \gtrsim (\Gamma + \frac{|eV|}{2})$, the amplitude of current moderately grows with the coupling $|\lambda|$ [see Fig. 2.17a, dotted lines]. The current maximum emerging at $\tilde{\varepsilon}_d = 0$ for $\beta \neq 0$ is reduced relative to its value at $\varepsilon_d = 0$ for $\beta = 0$. For finite QD-MBS hybridization ($|\lambda| \neq 0$) and $\beta \neq 0$, the current spectrum is visibly modified [see Fig. 2.17a, solid lines]. As in the EPI absence case, the suppression of current amplitude with the increase of $|\lambda|$ is observed when the renormalized QD level $\tilde{\varepsilon}_d$ is located within $-(\tilde{\Gamma} + \frac{|eV|}{2}) \lesssim \tilde{\varepsilon}_d \lesssim (\tilde{\Gamma} + \frac{|eV|}{2})$. Otherwise, when $|\tilde{\varepsilon}_d| \gtrsim (\tilde{\Gamma} + \frac{|eV|}{2})$, a grow in current amplitude with coupling $|\lambda|$ is seen. This kind of current behavior shows a qualitative agreement with the findings reported in [157] where the influence of EPI together with temperature T on couplings $|\lambda|$ and Γ is neglected by assuming those as constants. We discuss the influence of bias voltage eV on the $I - \tilde{\varepsilon}_d$ characteristics in the absence of EPI in Fig. 2.17b, and in the presence of EPI with coupling strength $\beta = 2.5\,\Gamma$ in Fig. 2.17c. According to

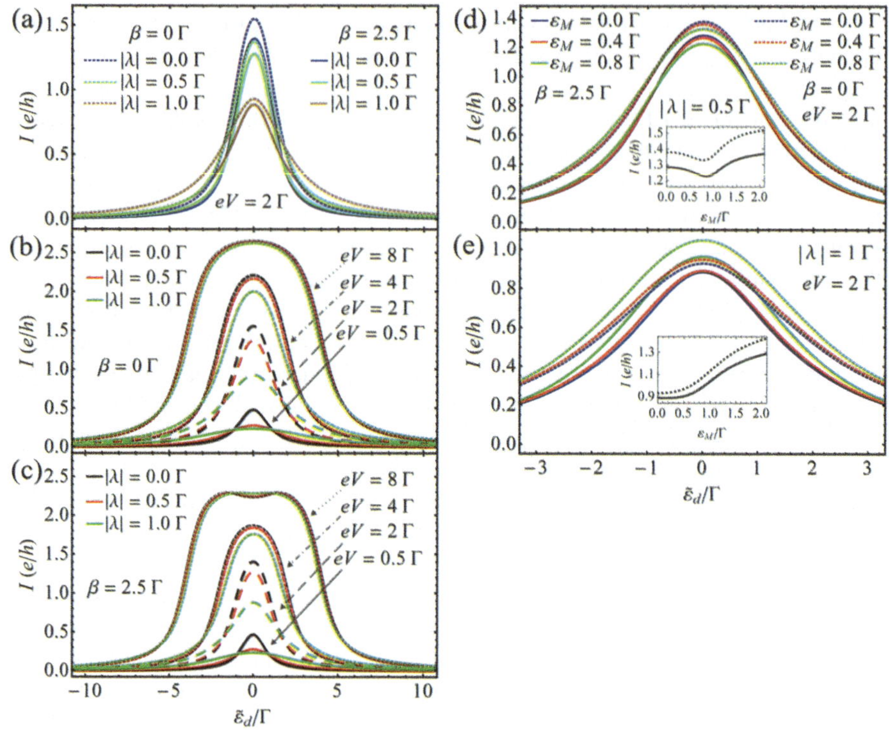

Fig. 2.17 **a** The current I versus $\tilde{\varepsilon}_d = \varepsilon_d - \beta^2/\omega_0$ with $|\lambda_1| = |\lambda|, |\lambda_2| = 0$ and $\phi = 0$ at $\varepsilon_M = 0\,\Gamma$ and $eV = 2\,\Gamma$. The dotted (solid) lines correspond to the $\beta = 0\,\Gamma$ ($\beta = 2.5\,\Gamma$) case. The current I versus $\tilde{\varepsilon}_d$ at different values of $|\lambda_1| = |\lambda|$ and bias voltage eV for $\varepsilon_M = 0\,\Gamma$, $|\lambda_2| = 0$ and $\phi = 0$ in the **b** absence and **c** presence of EPI with coupling strength $\beta = 2.5\,\Gamma$. The solid, dashed, dot-dashed and dotted lines represent the results corresponding to the bias voltages $eV = 0.5\,\Gamma$, $eV = 2\,\Gamma$, $eV = 4\,\Gamma$ and $eV = 8\,\Gamma$, respectively. The current I versus $\tilde{\varepsilon}_d = \varepsilon_d - \beta^2/\omega_0$ at different values of ε_M and $\lambda_1 = |\lambda|$ with $|\lambda_2| = 0$ and $\phi = 0$: **d** $|\lambda| = 0.5\,\Gamma$ and **e** $|\lambda| = 1\,\Gamma$. The dotted (solid) lines correspond to the $\beta = 0\,\Gamma$ ($\beta = 2.5\,\Gamma$) case. Insets show current I versus overlap energy ε_M at $\tilde{\varepsilon}_d = 0\,\Gamma$ under EPI with coupling strength $\beta = 2.5\,\Gamma$ (solid line), while at $\varepsilon_d = 0$ without EPI (dotted line) for: **d** $|\lambda| = 0.5\,\Gamma$ and **e** $|\lambda| = 1\,\Gamma$. The bias voltage in **d** and **e** is: $eV = 2\,\Gamma$. The temperature is $T = 0.1\,\Gamma$. Reproduced from [160] under CC-BY 4.0 license. Copyright 2023 by the authors

the literature [187], when $\beta = 0$ and $|\lambda| = 0$ [see Fig. 2.17b, black lines], the current magnitude increases with the grow in bias voltage $|eV|$. But, when the dot is attached to the MBS in the absence of EPI [see Fig. 2.17b, red and green lines], the amplitude of current $|I|$ is observably reduced compared to the $|\lambda| = 0$ case within the QD level domain $-(\Gamma + \frac{|eV|}{2}) \lesssim \varepsilon_d \lesssim (\Gamma + \frac{|eV|}{2})$, as found in Fig. 2.17a. When MZMs and EPI are introduced in the system [see Fig. 2.17c], the current I reacts similarly to the change in coupling $|\lambda|$ at low biases $|eV|$ as in the $\beta = 0$ limit under the mappings $\varepsilon_d \to \tilde{\varepsilon}_d$, and $\Gamma \to \tilde{\Gamma}$, respectively. Note that in the presence of EPI, the effect of dot-MZM hybridization $|\lambda|$ on the $I - \tilde{\varepsilon}_d$ characteristics alters depending on the value of bias voltage eV [see also Fig. 2.19a]. As a consequence, the change in magnitude of current due to the dot-MBS hybridization can be counteracted by tuning eV. We analyze the influence of Majorana overlap energy ε_M on the $I - \tilde{\varepsilon}_d$ characteristics when the dot connects to one MZM with coupling $|\lambda_1| = |\lambda|$ (and $|\lambda_2| = 0$ with $\phi = 0$). To do this, we plot, in Fig. 2.17d, e, the current I as a function of dot energy ε_d for $\beta = 0$, and as a function of renormalized dot energy $\tilde{\varepsilon}_d$ for $\beta = 2.5\,\Gamma$, at different Majorana overlap energy and dot-MZM coupling $|\lambda|$ values. The bias voltage and temperature are $eV = 2\,\Gamma$ and $T = 0.1\,\Gamma$, respectively. We see that the amplitude of current located at $\tilde{\varepsilon}_d = \varepsilon_d = 0$ in the absence of EPI and at $\tilde{\varepsilon}_d = 0$ in the presence of EPI reduces with the enhancement of overlap energy when the dot weakly hybridizes with the MZM [see Fig. 2.17d]. Therefore, we also observe that at a given overlap energy $\varepsilon_M \neq 0$, by moving on $\tilde{\varepsilon}_d$-axis away from $\tilde{\varepsilon}_d = 0$ with $\beta \neq 0$ (or $\varepsilon_d = 0$ with $\beta = 0$) to larger $|\tilde{\varepsilon}_d|$ (or $|\varepsilon_d|$) values, there exists a critical value above which I alters from a reduction to an amplification relative to the curve corresponding to $\varepsilon_M = 0$. The inset of Fig. 2.17d shows the current at $\tilde{\varepsilon}_d = 0$ with $\beta \neq 0$ (or $\varepsilon_d = 0$ with $\beta = 0$) as a function of ε_M. We see that the current amplitude at $\tilde{\varepsilon}_d = 0$ in the presence of EPI (or at $\varepsilon_d = 0$ in the absence of EPI) decreases with increasing ε_M and above an ε_M value a further increase in overlap energy leads to the current is starting to increase. For a stronger $|\lambda|$ value [see Fig. 2.17e with its inset], the current I responds differently to changes in overlap energy relative to the case of $|\lambda| = 0.5\,\Gamma$. Namely, at $\varepsilon_d = 0$ without EPI or at $\tilde{\varepsilon}_d = 0$ with EPI, the current magnitude $|I|$ grows with overlap energy ε_M. Note that when the dot level exceeds a critical energy value, the current amplitude at a given ε_M is reduced relative to the case of $\varepsilon_M = 0$ (not shown here). This critical QD energy value shifts to smaller $|\varepsilon_d|$ as ε_M increases. Note that the peak of the current curve is narrower for $\beta \neq 0$ [Fig. 2.17d, e, solid lines] than one for $\beta = 0$ [Fig. 2.17d, e, dashed lincs] due to renormalized dot-electrode coupling $\tilde{\Gamma}$.

To further investigate the regime $eV \neq 0$ and $|\lambda_1| \neq 0$ for $\varepsilon_M \neq 0$, we plot in Fig. 2.18a, b the current I as a function of QD-MZM coupling $|\lambda_1| = |\lambda|$ (and $|\lambda_2| = 0$ with $\phi = 0$) for different values of the Majorana overlap energy ε_M and bias voltage eV at $\varepsilon_d = 0$ in the absence of EPI and at $\tilde{\varepsilon}_d = 0$ in the presence of EPI with $\beta = 2.5\,\Gamma$, respectively, at a finite temperature $T = 0.1\,\Gamma$. We observe for degenerate MBSs with $\beta = 0$ [see Fig. 2.18a–b, blue dotted lines], that the current maximum arising at $\varepsilon_d = 0$ is remarkably reduced when the QD couples to the MZM, consistently to the results from Fig. 2.17a. We can see that the current amplitude $|I|$ increases with the bias $|eV|$ at fixed $|\lambda|$ values, in agreement with the findings from Fig. 2.17b. The value of dot-MZM coupling $|\lambda|$, where the current $|I|$ shows a significant reduction, moves to higher $|\lambda|$ values with the increase of $|eV|$. In the case of hybridized Majoranas, at a given ε_M value, the QD-MBS coupling presents a critical value, where the effect of overlap energy on the current magnitude changes from a reduction to an amplification relative to the $\varepsilon_M = 0$ curve. The critical $|\lambda|$ value depends on the bias eV. The growing bias voltage $|eV|$ results in negligible effect of the Majorana overlap energies ε_M, considered here, on the current characteristics. Note that a relation for current, $I(eV) = -I(-eV)$, can be established, as also seen in Fig. 2.19a–c. To have a complete idea of the bias voltage dependence of the current

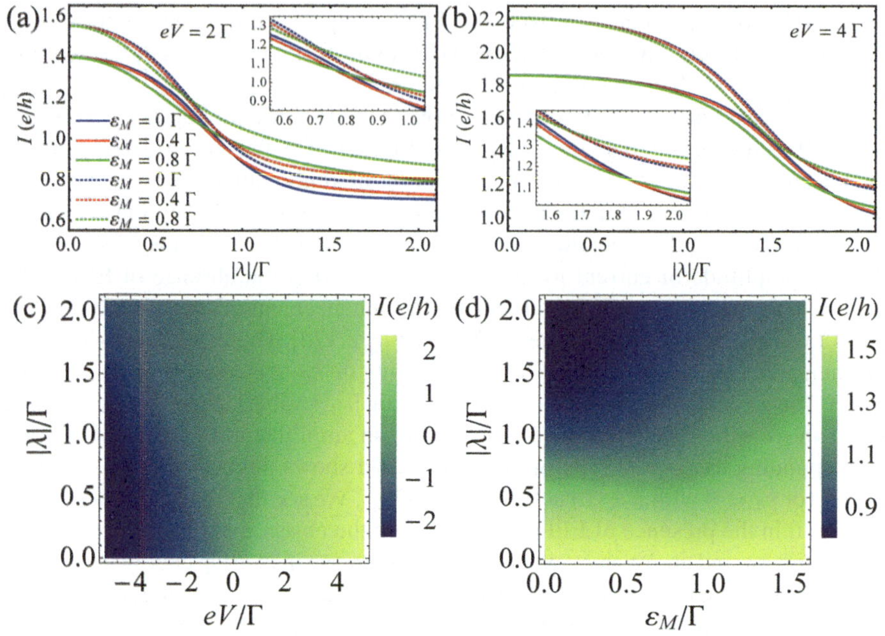

Fig. 2.18 The current I against the QD-MZM coupling $|\lambda_1| = |\lambda|$ (with $|\lambda_2| = 0$ and $\phi = 0$) for different overlap energy ε_M values at bias voltages: **a** $eV = 2\,\Gamma$ and **b** $eV = 4\,\Gamma$. The solid (dotted) lines correspond to the EPI presence case with $\beta = 2.5\,\Gamma$ at $\tilde{\varepsilon}_d = 0\,\Gamma$ (EPI absence case with $\beta = 0\,\Gamma$ at $\varepsilon_d = 0\,\Gamma$). **c** The current I versus bias voltage eV and dot-MBS coupling $|\lambda_1| = |\lambda|$ (with $|\lambda_2| = 0$ and $\phi = 0$) and $\varepsilon_M = 0\,\Gamma$ in the absence of EPI at $\varepsilon_d = 0\,\Gamma$. **d** The current I versus overlap energy ε_M and dot-MBS coupling $|\lambda_1| = |\lambda|$ (with $|\lambda_2| = 0$ and $\phi = 0$) at bias voltage $eV = 2\,\Gamma$ in the absence of EPI at $\varepsilon_d = 0\,\Gamma$. The temperature is $T = 0.1\,\Gamma$. Reprinted from [160] under CC-BY 4.0 license. Copyright 2023 by the authors

when the QD-MZM coupling varies, we plot in Fig. 2.18c the current I versus bias voltage eV and coupling $|\lambda_1| = |\lambda|$ (and $|\lambda_2| = 0$ with $\phi = 0$) for degenerate MBSs at dot level $\varepsilon_d = 0$ and temperature $T = 0.1\,\Gamma$ in the absence of EPI. We note the line $|\lambda| \propto eV$ in the current map represents an inflection point corresponding to a peak in the differential conductance. In Fig. 2.18d, the current I as a function of Majorana overlap energy ε_M and dot-MZM coupling strength $|\lambda_1| = |\lambda|$ (and $|\lambda_2| = 0$ with $\phi = 0$) is plotted at bias voltage $eV = 2\,\Gamma$, QD energy $\varepsilon_d = 0$ and temperature $T = 0.1\,\Gamma$ in the absence of EPI. We observe that the current I map details further changes with the increase of ε_M when the dot is strongly connected to the MZM [see also Fig. 2.18a]. For $\beta \neq 0$ [see Fig. 2.18a, b, solid lines], the current magnitude $|I|$ is decreased relative to the EPI absence case, which is more observable at voltages around $eV \approx 4\,\Gamma$ [see also Fig. 2.19a–c for a larger bias voltage domain]. In the presence of EPI the $I - |\lambda|$ curves exhibit the same behavior, as those in the $\beta = 0$ limit at low bias voltages.

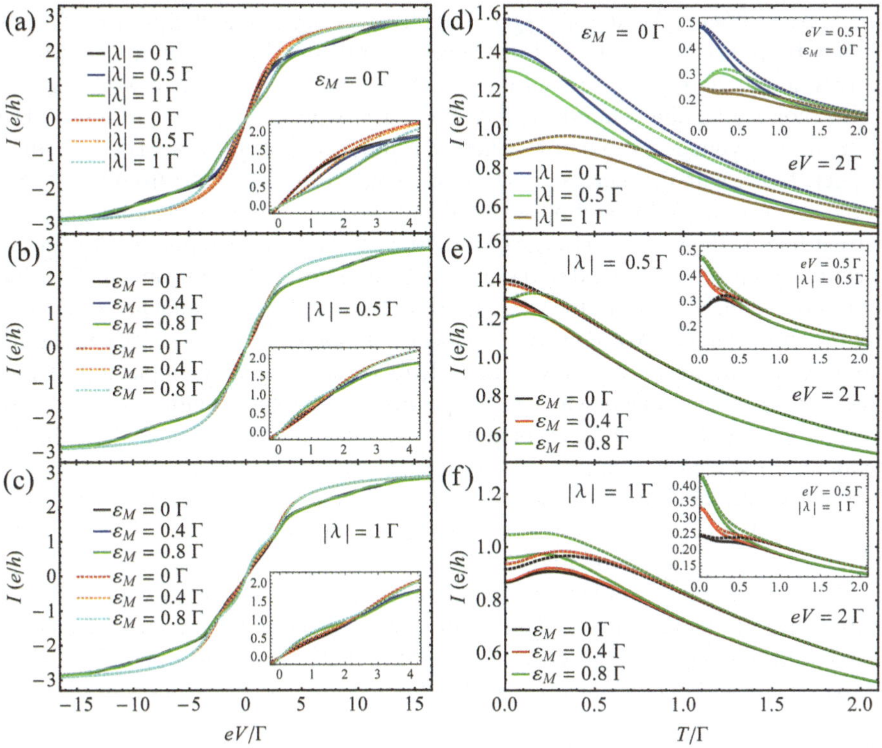

Fig. 2.19 **a** The current I versus bias voltage eV at different values of $|\lambda_1| = |\lambda|$ for overlap energy $\varepsilon_M = 0\,\Gamma$. The current I versus bias voltage eV at different values of ε_M and coupling $|\lambda|$: **b** $|\lambda| = 0.5\,\Gamma$ and **c** $|\lambda| = 1\,\Gamma$, respectively. The temperature is $T = 0.1\,\Gamma$. The insets in each panel zoom in on I at low positive voltages. **d** The current I versus temperature T at different values of $|\lambda_1| = |\lambda|$ for $\varepsilon_M = 0\,\Gamma$ at $eV = 2\,\Gamma$. The inset in **d** represents the results at $eV = 0.5\,\Gamma$. The current I versus temperature T at different ε_M values at $eV = 2\,\Gamma$ for: **e** $|\lambda_1| = |\lambda| = 0.5\,\Gamma$ and **f** $|\lambda_1| = |\lambda| = 1\,\Gamma$, respectively. The insets in **e** and **f** represent the results at $eV = 0.5\,\Gamma$. In all cases the solid (dotted) lines show the results for $\beta = 2.5\,\Gamma$ at $\tilde{\varepsilon}_d = 0\,\Gamma$ ($\beta = 0\,\Gamma$ at $\varepsilon_d = 0\,\Gamma$) with $|\lambda_2| = 0$ and $\phi = 0$. Reproduced from [160] under CC-BY 4.0 license. Copyright 2023 by the authors

In what follows, we analyze the current–bias voltage ($I - eV$) characteristics of a system where the QD hybridizes with one MZM located at one of the sites of a TSNW ($|\lambda_1| = |\lambda|$ and $|\lambda_2| = 0$ with $\phi = 0$) at different values of the dot-MZM coupling $|\lambda|$ for both unhybridized and hybridized Majoranas, in the absence and presence of EPI. The results are illustrated in Fig. 2.19a–c for a finite temperature $T = 0.1\,\Gamma$ at $\varepsilon_d = 0$ without EPI and at $\tilde{\varepsilon}_d = 0$ with EPI of coupling strength $\beta = 2.5\,\Gamma$. We observe that the $I - eV$ characteristics shows a step-like structure which is explained below. In the absence of Majoranas and EPI, when the QD is positively biased ($eV > 0$), and $\mu_R < \varepsilon_d < \mu_L$ (by considering $\mu_L = -\mu_R = eV/2$), the QD is able to receive an electron from left (L) electrode and to transfer it to right (R) electrode, which leads to a current flowing through the QD. Such a QD setup (without EPI or MBS) has already been discussed in Ref. [188]. When the QD hybridizes with the MZM, the current amplitude reduces around the zero-bias voltage with the grow of coupling $|\lambda|$ for nonoverlapping Majoranas [see Fig. 2.19a with its inset, dotted lines]. The width of bias voltage eV window, where I is influenced by the dot-MZM coupling $|\lambda|$, gets larger with the increase in $|\lambda|$, consistently with the results from Fig. 2.18c. In the absence of MZMs and presence of EPI [see Fig. 2.19a, black solid line], the current-bias voltage curve is clearly modified, because new steps develop in the spectrum corresponding to the opening of a series of additional phonon-assisted transport channels [189]. In the presence of EPI when the dot is attached to the one of the nonoverlapping MZMs ($\varepsilon_M = 0$) located at opposite sites of a TSNW, further modifications in the current-bias voltage characteristics can be seen [see Fig. 2.19a, blue and green solid lines]. Thus, in the positive bias voltage region (i.e., $eV \gtrsim 0$), we can observe that the current magnitude reduces with the increase of dot-MZM coupling $|\lambda|$ when eV is approximately located within $2p\omega_0 \lesssim eV \lesssim (2p+1)\omega_0$. In contrast, the amplitude of current increases with the coupling $|\lambda|$ in the bias voltage regimes $(2p+1)\omega_0 \lesssim eV \lesssim 2(p+1)\omega_0$ with $p = 0, 1, 2, ...$ These results are in agreement with the findings of Ref. [157]. At negative bias voltages ($eV \lesssim 0$), the current reacts similarly to the changes in $|\lambda|$ as in the case $eV \gtrsim 0$. By considering the antisymmetric nature of the current–bias voltage curves, we find that the current magnitude $|I|$ decreases with the enhancement of coupling $|\lambda|$ when the bias eV is within $(2p+1)\omega_0 \lesssim eV \lesssim 2(p+1)\omega_0$, and $|I|$ grows with coupling $|\lambda|$ when eV is situated within $2p\omega_0 \lesssim eV \lesssim (2p+1)\omega_0$ with $p = -1, -2, ...,$ respectively. We now illustrate in Fig. 2.19b, c the influence of Majorana overlap energy on the $I - eV$ characteristics at two different values of the QD-MZM coupling $|\lambda|$, both in the absence of EPI at $\varepsilon_d = 0$, and in the presence of EPI of coupling strength $\beta = 2.5\,\Gamma$ at $\tilde{\varepsilon}_d = 0$. In the $\beta = 0$ case [see Fig. 2.19b, c, dotted lines], we see that the magnitude of current $|I|$ grows with the Majorana overlap energy ε_M around the zero-bias region, when the bias eV is within $|eV| \lesssim \mathcal{V}$. Here, we introduced $\mathcal{V} \propto (2|\lambda| + \varepsilon_M)$ with $|\lambda| \neq 0$. Otherwise, when the voltage is $|eV| \gtrsim \mathcal{V}$, the current magnitude $|I|$ slightly decreases with the enhancement in ε_M. In the $\beta \neq 0$ case with hybridized MZMs [see Fig. 2.19b, c, solid lines], the finite overlap energy observably affects the current–bias voltage characteristics. If the QD setup is positively biased (i.e., $eV \gtrsim 0$), the current magnitude $|I|$ enhances with Majorana overlap energy ε_M at voltages within $2p\omega_0 \lesssim eV \lesssim 2p\omega_0 + \tilde{\mathcal{V}}$ and $(2p+1)\omega_0 \lesssim eV \lesssim 2(p+1)\omega_0 - \tilde{\mathcal{V}}$. While $|I|$ reduces with increasing ε_M when the bias eV is within $2p\omega_0 + \tilde{\mathcal{V}} \lesssim eV \lesssim (2p+1)\omega_0$ and $2(p+1)\omega_0 - \tilde{\mathcal{V}} \lesssim eV \lesssim 2(p+1)\omega_0$, with $p = 0, 1, 2, ...$ Here, $\tilde{\mathcal{V}} \propto (2|\tilde{\lambda}| + \varepsilon_M)$ for $|\tilde{\lambda}| \neq 0$ with $0 < \tilde{\mathcal{V}} < \omega_0$. In the case of a negatively biased dot (i.e., $eV \lesssim 0$), the alteration of the response of current to the change in overlap energy ε_M can be expressed similarly to the $eV \gtrsim 0$ case by considering the asymmetric property of the current–bias voltage characteristics. As a consequence, when the QD hybridizes with the MZM, the current $|I|$ can be reduced or amplified by varying the Majorana overlap energy ε_M at a given value of the bias voltage eV.

We plot in Fig. 2.19d–f the current I versus temperature T in the case when the dot hybridizes with one MZM in the TSNW ($|\lambda_1| = |\lambda|$ and $|\lambda_2| = 0$ with $\phi = 0$)

for different values of coupling $|\lambda|$, Majorana overlap energy ε_M and bias voltage eV. The results are calculated at $\varepsilon_d = 0$ in the absence of EPI and at $\tilde{\varepsilon}_d = 0$ in the presence of EPI with coupling strength $\beta = 2.5\,\Gamma$, respectively. We can see that the current amplitude I is suppressed in the presence of EPI relative to its value in the $\beta = 0$ limit [see Fig. 2.19d–f and their inset, solid and dotted lines]. One observes that I presents a nonmonotonic behavior with temperature for nonoverlapping MBSs when the dot weakly connects to the TSNW at low voltages eV for both $\beta = 0$ and $\beta \neq 0$ cases [see inset of Fig. 2.19d, green dotted and solid lines]. Namely, the current magnitude grows with temperature up to a specific value and starts reducing above that value. With the increase of coupling $|\lambda|$ this behavior of the current–temperature characteristics begins to disappear [see inset of Fig. 2.19d, brown dotted and solid lines]. With the increase of bias voltage, the nonmonotonic behavior of I with temperature appears at stronger QD-MBS coupling [Fig. 2.19d, brown dotted and solid lines]. We notice that this tendency of the current–temperature characteristics vanishes at higher bias voltages eV with values of the coupling $|\lambda|$ in accordance with the approximation employed here. For hybridized Majoranas with relatively strong ε_M, the nonmonotonic behavior of I with T vanishes at small bias voltages [see the inset of Fig. 2.19e, red and green lines]. Additionally, the current $|I|$ changes nonmonotonically with T for weaker dot-MZM coupling $|\lambda|$ with stronger overlap energy ε_M at bias voltage $eV = 2\,\Gamma$ [see Fig. 2.19e, green dotted and solid lines]. This behavior of the current–temperature curves begins to disappear with further increase of the overlap energy ε_M (not shown here). In the case when the dot strongly hybridizes with the MZM, the current $|I|$ reduces with the increase of temperature T at low bias voltages [see the inset of Fig. 2.19f at $eV = 0.5\,\Gamma$]. The nonmonotonic behavior of the $I - T$ curves is present for values of overlap energy ε_M considered here at bias voltage $eV = 2\,\Gamma$ with QD-MBS coupling $|\lambda| = 1\,\Gamma$ [see Fig. 2.19f]. As in the case of $|\lambda| = 0.5\,\Gamma$, a further grow in overlap energy results in smearing of the nonmonotonic behavior of $I - T$ curves at bias $eV = 2\,\Gamma$ (not shown here). Note also that the nonmonotonic tendency of the current–temperature characteristics disappears at higher bias voltages eV independently of the values of QD-MBS coupling $|\lambda|$ within the limits defined by the approximations applied here. In this case, the Majorana overlap energy ε_M presents less impact on the current–temperature characteristics. As a consequence, the response of I to the change in T is nontrivially altered depending on the dot-MZM coupling $|\lambda|$, Majorana overlap energy ε_M and bias voltage eV.

In the following, we study the linear conductance $\mathcal{G}$ at zero and finite temperatures in the presence of EPI. To do this, we plot in Fig. 2.20 the linear conductance $\mathcal{G}$ versus magnetic flux phase ϕ at zero and a finite temperature for QD energy $\varepsilon_d = 0$, in the cases $\beta = 0$ and $\beta \neq 0$, for unhybridized and hybridized MBSs. As presented in Sect. 2.3.1 (see Figs. 2.8 and 2.9), without EPI ($\beta = 0$) for degenerate MBSs, the zero-temperature linear conductance $\mathcal{G}$ becomes ε_d-independent and shows a 2π periodicity against the magnetic flux phase ϕ. In $\beta \neq 0$ with $\varepsilon_M = 0$ case, the approximated form of dot retarded Green's functions in the limit $\varepsilon \to 0$, i.e., $\tilde{G}^r_{d11}(\varepsilon \to 0)$ and $\tilde{G}^r_{d12}(\varepsilon \to 0)$, are given by Eq. (2.33) under mappings $\Gamma \to \tilde{\Gamma}$ and

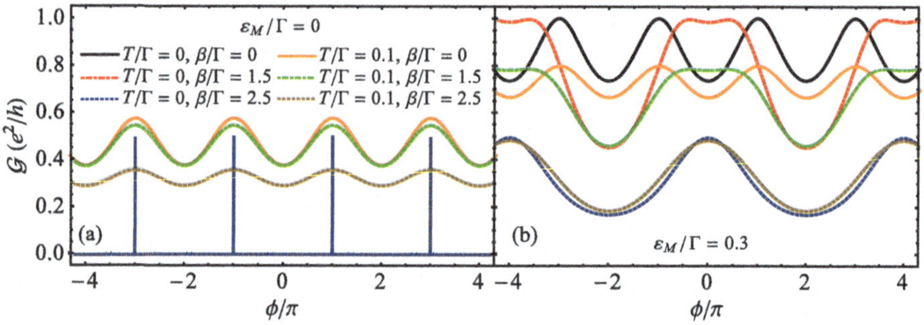

Fig. 2.20 The variation of linear conductance $\mathcal{G}$ with magnetic flux phase ϕ at different values of the electron-phonon coupling strength β for: **a** nonoverlapping $\varepsilon_M = 0\,\Gamma$ and **b** overlapping MBSs $\varepsilon_M = 0.3\,\Gamma$. The QD is symmetrically attached to both MZMs with coupling strengths $|\lambda_1| = |\lambda_2| = 0.3\,\Gamma$ at dot level $\varepsilon_d = 0\,\Gamma$, and zero or a finite $T = 0.1\,\Gamma$ temperature. Reproduced with permission from [159]. Copyright 2022, American Physical Society

$|\lambda_j| \rightarrow |\tilde{\lambda}_j|$. Similarly to the EPI absence case, these Green's functions are independent of the renormalized QD energy $\tilde{\varepsilon}_d$ and finite values of the dot-MZM couplings $|\tilde{\lambda}_j|$. Therefore, the zero-temperature linear conductance does not depend on the electron-phonon coupling strength β for nonoverlapping MBSs. This behavior of $\mathcal{G}$ was also reported in a device where a QD is attached to one of the ends of a TSNW in the presence of EPI [156]. The linear conductance preserves its 2π periodicity in the presence of EPI for unhybridized Majoranas. As we have seen above, for $\beta = 0$, the linear conductance $\mathcal{G}$ has a 4π periodicity as a function of ϕ when the MZMs hybridize ($\varepsilon_M \neq 0$) and the QD level is tuned, i.e., $\varepsilon_d \neq 0$. In the $\beta \neq 0$ case with hybridized MBSs, $\mathcal{G}$ depends on the QD level and electron-phonon coupling strength both at zero and nonzero temperatures. For hybridized MZMs, the 2π periodicity of linear conductance as a function of ϕ changes to 4π under EPI which is attributed to the finite renormalized QD energy $\tilde{\varepsilon}_d$.

Figure 2.21 shows the linear conductance $\mathcal{G}$ versus magnetic flux phase ϕ and Majorana overlap energy ε_M for symmetrical dot-MZM couplings $|\lambda_1| = |\lambda_2|$ at zero temperature and QD energy $\varepsilon_d = 0$ in the absence and presence of EPI. One sees from Fig. 2.21a that the zero-temperature linear conductance $\mathcal{G}$ presents a 2π periodicity in magnetic flux phase ϕ when $\beta = 0$, independent of ε_M values, in agreement with the findings of Fig. 2.5a–c. In the case of strong MZM-MZM coupling ε_M, the minima of zero-temperature linear conductance $\mathcal{G}$ are remarkably increased while conserving its 2π periodicity. In the presence of EPI [see Fig. 2.21b], the periodicity of linear conductance $\mathcal{G}$ as a function of ϕ switches from 2π to 4π provided that the Majorana overlap energy ε_M is finite. For unhybridized MBSs, the linear conductance $\mathcal{G}$ has 2π periodicity showing minima at magnetic flux phases $\phi \neq (2n+1)\pi$, with n being an integer. Therefore, $\mathcal{G}$ reaches the value $e^2/2h$ at $\phi = (2n+1)\pi$, independent of β. For finite Majorana overlap energy values $\varepsilon_M > 0$, the linear conductance shows a 4π periodicity against ϕ. At magnetic flux phase $\phi = 4n\pi$, the linear conductance contains local minima for small ε_M values, but for larger overlap energies, $\mathcal{G}$ has

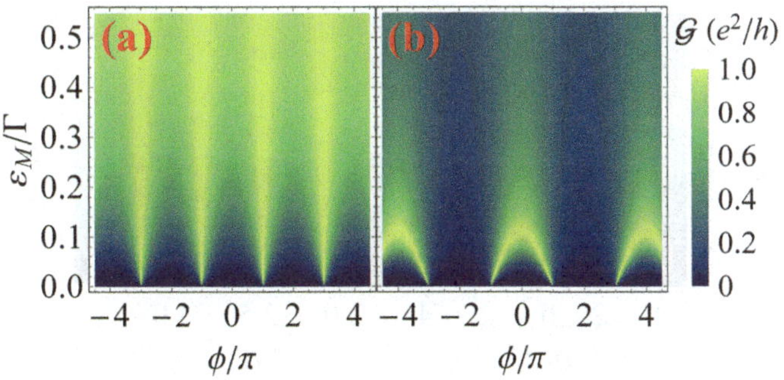

Fig. 2.21 The linear conductance $\mathcal{G}$ against the magnetic flux phase ϕ and Majorana overlap energy ε_M at zero temperature for symmetrically coupled QD-MBSs with $|\lambda_1| = |\lambda_2| = 0.3\,\Gamma$, at dot energy $\varepsilon_d = 0\,\Gamma$, in the **a** absence ($\beta = 0\,\Gamma$) and **b** presence of EPI with coupling strength $\beta = 2.5\,\Gamma$. Reprinted with permission from [159]. Copyright 2022, American Physical Society

maxima. A further increase in ε_M, results in the reduction of $\mathcal{G}$ peaks. Consequently, the linear conductance periodicity goes through a transition from 2π at $\varepsilon_M = 0$ to 4π at overlap energies $\varepsilon_M > 0$ in the presence of EPI.

We now analyze the influence of QD level on the linear conductance. Figure 2.22a shows the results for linear conductance $\mathcal{G}$ versus magnetic flux phase ϕ and QD energy ε_d in the presence of EPI at zero-temperature for hybridized MBSs when the dot is attached to both MZMs with symmetrical couplings $|\lambda_1| = |\lambda_2|$. The features of the linear conductance are clearly recognized by recalling the fact that in a QD-MBSs system without EPI, $\mathcal{G}$ presents a 2π periodicity in flux phase ϕ when the QD level is not tuned ($\varepsilon_d = 0$). This 2π periodicity transforms to 4π by tuning the dot energy ($\varepsilon_d \neq 0$) for overlapping MBSs ($\varepsilon_M \neq 0$, see Fig. 2.9). In the presence of EPI [see Fig. 2.22a], the QD level is renormalized, i.e., $\varepsilon_d \to \tilde{\varepsilon}_d = \varepsilon_d - \beta^2/\omega_0$, that results in a shift in the map of linear conductance $\mathcal{G}$ in the positive direction on dot energy ε_d-axis. Therefore, at dot level $\varepsilon_d = \beta^2/\omega_0$, the linear conductance $\mathcal{G}$ presents a 2π-periodicity as a function of ϕ. Otherwise, when $\varepsilon_d \neq \beta^2/\omega_0$, $\mathcal{G}$ shows a 4π periodicity in magnetic flux phase ϕ. Figure 2.22b illustrates the results for linear conductance $\mathcal{G}$ versus electron-phonon coupling strength β and dot level ε_d at zero temperature and magnetic flux phase $\phi = \pi$ for hybridized MBSs ($\varepsilon_M \neq 0$) and symmetrical dot-MZM couplings $|\lambda_1| = |\lambda_2|$. We note that the 2π-periodicity of linear conductance is found only if the renormalized QD energy $\tilde{\varepsilon}_d$ disappears, which requires the condition $\varepsilon_d > 0$, because of $\beta^2/\omega_0 > 0$. Figure 2.22c shows the zero-temperature linear conductance $\mathcal{G}$ versus magnetic flux phase ϕ at different QD energy ε_d values, in the absence of EPI and in the presence of EPI with coupling strength $\beta = 2.5\,\Gamma$ for hybridized MBSs. It is clearly seen that the influence of EPI on linear conductance can be eliminated with a good approximation at $\tilde{\varepsilon}_d = 0$ when tuning the QD level to $\varepsilon_d = \beta^2/\omega_0$. Remarkably, the oscillations in $\mathcal{G}$ for two opposite values of the renormalized QD energy $\tilde{\varepsilon}_d$ are in antiphase as can be recognized also in Fig. 2.9 for $\beta = 0$. In summary, the most important message of Fig. 2.22 is that when the EPI is switched on, the periodicity of linear conductance $\mathcal{G}$ as a function

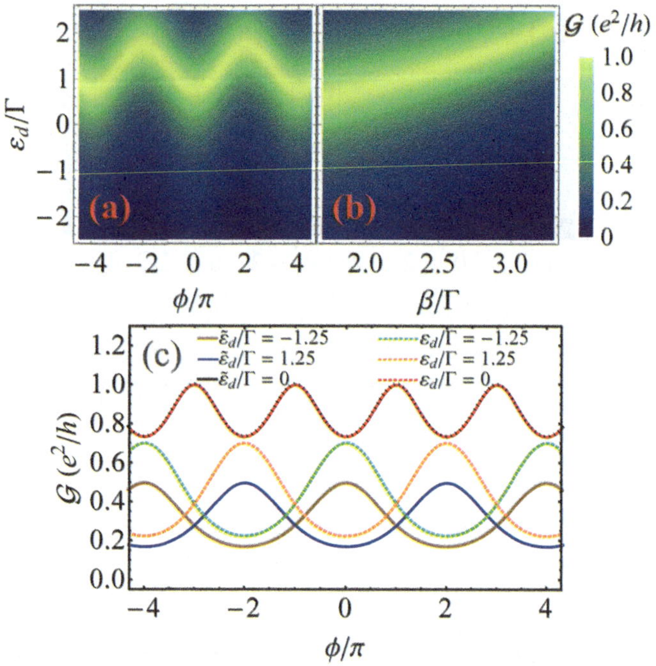

Fig. 2.22 **a** The variation of linear conductance $\mathcal{G}$ with magnetic flux phase ϕ and dot energy ε_d at zero temperature in the presence of EPI with coupling strength $\beta = 2.5\,\Gamma$. **b** $\mathcal{G}$ as a function of β and ε_d at $\phi = \pi$ and zero temperature. **c** $\mathcal{G}$ versus magnetic flux phase ϕ at different ε_d values, in the absence and presence of EPI where the renormalized dot energy is $\tilde{\varepsilon}_d = \varepsilon_d - \beta^2/\omega_0$ with EPI strength $\beta = 2.5\,\Gamma$. The solid (dashed) lines correspond to the $\beta = 2.5\,\Gamma$ ($\beta = 0\,\Gamma$) case. The rest of parameters are chosen as: the Majorana overlap energy $\varepsilon_M = 0.3\,\Gamma$ and the symmetrical QD-MBS couplings $|\lambda_1| = |\lambda_2| = 0.3\,\Gamma$. Reprinted with permission from [159]. Copyright 2022, American Physical Society

of ϕ switches between 2π and 4π values when changing $\tilde{\varepsilon}_d$ through the dot level ε_d or EPI strength β for hybridized MBSs ($\varepsilon_M \neq 0$).

In what follows, we analyze the effect of temperature on linear conductance $\mathcal{G}$ in the presence of EPI. The linear conductance generally depends strongly on temperature since $\mathcal{G} \propto T^{-1} \int d\varepsilon\, \mathcal{A}_d(\varepsilon) f(\varepsilon)[1 - f(\varepsilon)]$, which leads to a decay with an overall prefactor T^{-1}. The results presented here in Fig. 2.23 expand on the findings from Fig. 2.20 where only a single finite value for the temperature was assumed. Figure 2.23 illustrates the linear conductance $\mathcal{G}$ versus magnetic flux phase ϕ and renormalized dot energy $\tilde{\varepsilon}_d$, with EPI, at different values of temperature, for $\varepsilon_M = 0$ and $\varepsilon_M \neq 0$, when the dot connects to both MBSs with fixed symmetrical dot-Majorana couplings $|\lambda_j| = |\lambda|$. Similarly to the findings from Fig. 2.11, we observe, in the presence of EPI, that the increasing T leads to a broadening of the Dirac-comb peak structure which shows up at zero temperature for unhybridized MBSs. Beside the enhancement of minima with the increase in temperature T at weak values of the symmetrical coupling strength $|\lambda|$, an amplification of maxima in linear conductance spectrum is seen

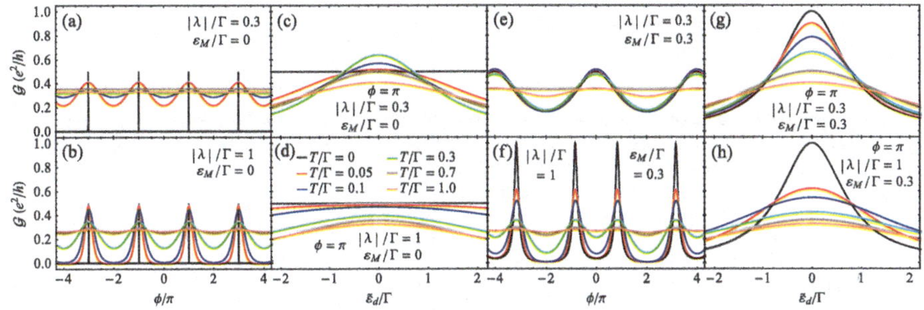

Fig. 2.23 **a, b, e, f** The linear conductance $\mathcal{G}$ versus magnetic flux phase ϕ at QD energy $\varepsilon_d = 0\,\Gamma$, and **c, d, g, h** $\mathcal{G}$ versus renormalized dot energy $\tilde{\varepsilon}_d$ for magnetic flux phase $\phi = \pi$, at different temperatures T with EPI. The QD couples with symmetrical couplings $|\lambda_j| = |\lambda|$ to both the ends of a TSNW. The electron-phonon coupling strength is $\beta = 2.5\,\Gamma$. The Majorana overlap energy ε_M is: **a–d** $\varepsilon_M = 0\,\Gamma$ and **e–h** $\varepsilon_M = 0.3\,\Gamma$. The symmetrical dot-MZM coupling strength $|\lambda|$ is: **a, c, e, g** $|\lambda| = 0.3\,\Gamma$ and **b, d, f, h** $|\lambda| = 1\,\Gamma$. Reprinted with permission from [159]. Copyright 2022, American Physical Society

at renormalized dot energy $\tilde{\varepsilon}_d = 0$ [Fig. 2.23a, c]. The grow in temperature leads to the linear conductance exceeds the $e^2/2h$ value for unhybridized MBSs, after which, the conductance drops below the value $e^2/2h$ at higher temperatures. Accordingly to the findings for $\beta = 0$ from Fig. 2.11, the linear conductance peaks are suppressed at higher temperatures in the presence of EPI. In addition, we find that for stronger QD-MBS couplings $|\lambda|$, the peaks of linear conductance reduce monotonically with temperature from the value $e^2/2h$ at zero temperature and $\tilde{\varepsilon}_d = 0$ [Fig. 2.23b, d]. The width of resonances is narrower compared to the weak $|\lambda|$ coupling case. The resonances are flattened out to the point of masking the conductance periodicity at high temperatures. In the case of hybridized MBSs ($\varepsilon_M \neq 0$), $\mathcal{G}$ exhibits a 4π periodic function in ϕ due to the finite renormalized QD energy $\tilde{\varepsilon}_d$, independent of the finite values of coupling $|\lambda|$ [Fig. 2.23e, f]. The conductance peaks become sharper as the coupling $|\lambda|$ gets stronger. The maxima of linear conductance reach e^2/h at $\tilde{\varepsilon}_d = 0$ and $T = 0$, after that they decrease with growing T [see Fig. 2.23g, h]. Also note that the maxima of linear conductance do not surpass the e^2/h value with the grow of temperature as was reported in the no EPI case (Fig. 2.11).

Now, we discuss the magnetic flux Φ dependence of differential conductance dI/dV in the presence of EPI. In Fig. 2.24a, b the differential conductance dI/dV versus magnetic flux phase ϕ is plotted at zero temperature for degenerate and hybridized Majoranas at different electron-phonon coupling strength β values. The dot connects to both MZMs of a TSNW with weak symmetrical couplings $|\lambda_1| = |\lambda_2|$ at QD energy $\varepsilon_d = 0$ and a given value of bias voltage. The differential conductance shows a 2π periodicity in magnetic flux phase ϕ for unhybridized MZMs ($\varepsilon_M = 0$), in the case of $\beta = 0$ [Fig. 2.5d–f]. When $\beta \neq 0$, the minima of zero-temperature differential conductance dI/dV, originally located at $\phi = 2n\pi$ in the absence of EPI (n is an integer), start to grow with β. When the value of EPI coupling strength β approaches $\beta \approx 1.685\,\Gamma$ [see Fig. 2.24a, blue line], then the local minima of dI/dV

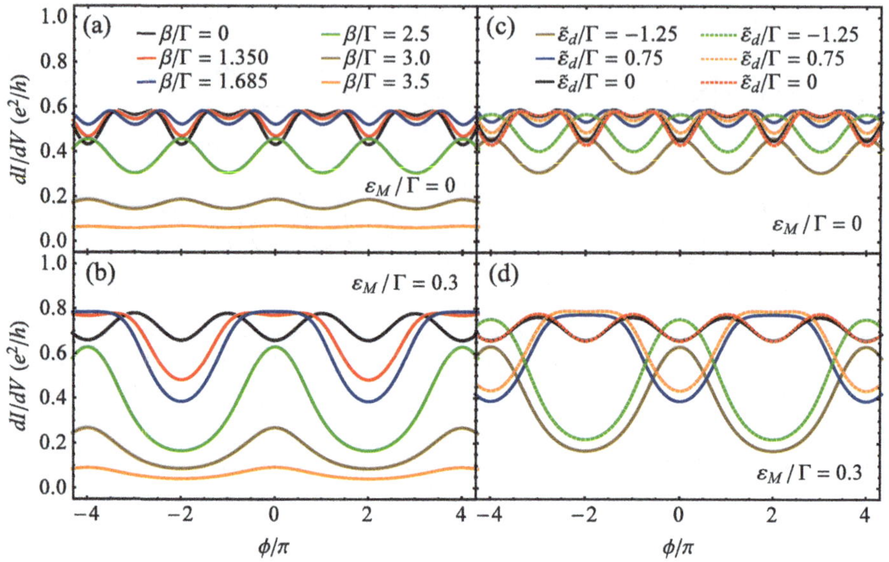

Fig. 2.24 **a**, **b** The differential conductance dI/dV versus magnetic flux phase ϕ for different values of the EPI strength β at dot energy $\varepsilon_d = 0\,\Gamma$. **c**, **d** The differential conductance dI/dV versus magnetic flux phase ϕ at different dot energies ε_d, with and without EPI ($\tilde{\varepsilon}_d = \varepsilon_d - \beta^2/\omega_0$, $\beta = 2.5\,\Gamma$), where the solid (dashed) curves represent the results in the presence (absence) of EPI. The rest of parameters are chosen as: bias voltage $eV = 0.28\,\Gamma$, dot-MZM couplings $|\lambda_1| = |\lambda_2| = 0.3\,\Gamma$ and zero temperature. The Majorana overlap energy is: **a**, **c** $\varepsilon_M = 0\,\Gamma$ and **b**, **d** $\varepsilon_M = 0.3\,\Gamma$. Reproduced with permission from [159]. Copyright 2022, American Physical Society

showing up at $\phi = (2n+1)\pi$ are identical with those of emerging at $\phi = 2n\pi$ resulting in the differential conductance becomes approximately π-periodic. A further enhancement in electron-phonon coupling β leads to the recovering the 2π periodicity of dI/dV in ϕ, with maxima occurring now at $\phi = 2n\pi$. The dI/dV curve is totally suppressed at stronger EPI coupling strengths β. When the MBSs overlap ($\varepsilon_M \neq 0$) [see Fig. 2.24b], the zero-temperature differential conductance dI/dV begins with a 2π periodic dependence on ϕ with minima emerging at $\phi = 2n\pi$ for $\beta = 0$, which transforms to 4π in the presence of EPI. The minima of dI/dV, initially showing up at $\phi = 4n\pi$ for $\beta = 0$, change to local plateaus when β approaches a critical value, namely, $\beta \approx 1.5\,\Gamma$. A further increase in EPI coupling strength β, within the limits of approximation employed here, leads to the transformation of conductance plateaus to maxima and to those suppression at larger β. Briefly, under variation of β, we can achieve a π-phase shift in the differential conductance dI/dV for nonoverlapping and a transition in its periodicity between 2π and 4π for overlapping MBSs. Figure 2.24c, d show the results for differential conductance dI/dV versus magnetic flux phase ϕ at zero temperature, in the absence and presence of EPI, when the dot level is tuned $\varepsilon_d \neq 0$, for nonoverlapping and overlapping Majorana wavefunctions. The QD connects to both MBSs with symmetrical couplings

$|\lambda_1| = |\lambda_2|$ when β and eV are fixed. We can see that, for electron-phonon coupling strength $\beta = 2.5\,\Gamma$, dI/dV exhibits a 2π-periodicity as a function of ϕ at $\tilde{\varepsilon}_d = 0$, independent of the ε_M values, presenting minima at $\phi = 2n\pi$ [see Fig. 2.24c, d]. In the case when ε_M and $\tilde{\varepsilon}_d$ are finite [see Fig. 2.24d], the periodicity of dI/dV is 4π in magnetic flux phase ϕ. When the dot level meets $\varepsilon_d = \beta^2/\omega_0$, the renormalized QD level disappears $\tilde{\varepsilon}_d = 0$, resulting in the suppression of the effect of EPI on the differential conductance periodicity. Therefore, the periodicity of dI/dV as a function of magnetic flux phase approaches the $\beta = 0$ case with dot energy $\varepsilon_d = 0$. Thus, the periodicity of differential conductance alters when tuning the QD energy level ε_d.

Finally, we examine in Fig. 2.25 the variation of differential conductance dI/dV with bias voltage eV, in the absence and the presence of EPI, when the dot weakly connects to both MZMs, located in the opposite sites of a TSNW, with symmetrical couplings $|\lambda_1| = |\lambda_2| = |\lambda|$, at zero and finite temperatures. We consider nonoverlapping and overlapping MBSs and different values of the magnetic flux phase ϕ when the dot level is not tuned $\varepsilon_d = 0$. We find in the $\beta = 0$ limit at low temperatures with weak coupling $|\lambda|$ that three resonances develop in the differential conductance spectra for hybridized MBSs ($\varepsilon_M \neq 0$) [see Fig. 2.25b, d, dashed lines], and two peaks show up for unhybridized Majoranas [see Fig. 2.25a, c, dashed lines]. Specifically, a

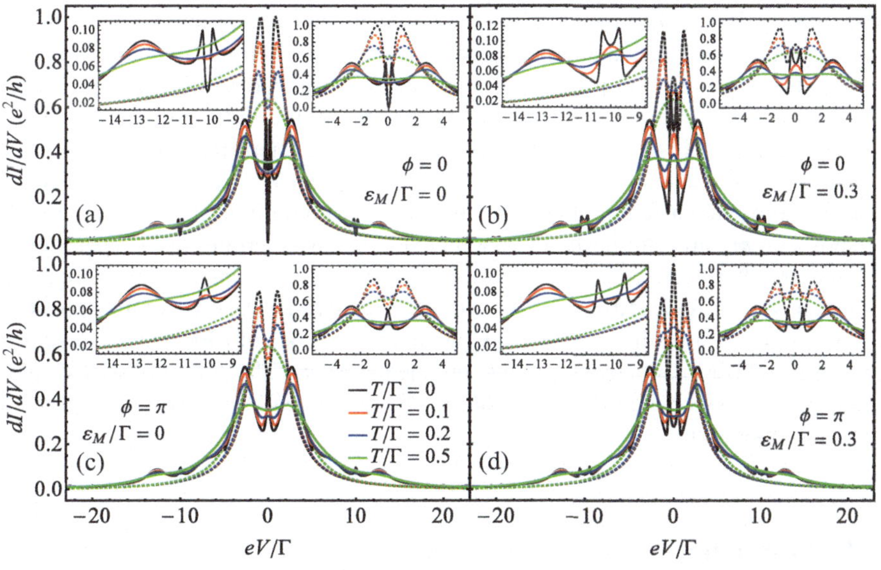

Fig. 2.25 The variation of differential conductance dI/dV with bias voltage eV at different temperatures T in the presence of EPI when the dot is attached to both MZMs with symmetrical couplings $|\lambda_1| = |\lambda_2| = 0.3\,\Gamma$ for two magnetic flux phases ϕ. The dot energy is $\varepsilon_d = 0\,\Gamma$. The solid (dashed) lines correspond to the $\beta = 2.5\,\Gamma$ ($\beta = 0\,\Gamma$) case. The rest of the parameters are presented in the figure. The right and left insets in each panel are zoom in on the central and first phonon-assisted peak in negative energy domain, respectively. Reprinted with permission from [159]. Copyright 2022, American Physical Society

resonance occurs for $\varepsilon_M \neq 0$ at $eV = 0$ in the differential conductance, while a minimum develops at zero bias for $\varepsilon_M = 0$. With the increase in temperature, the amplitude of resonant peaks decreases. The central resonance located at $eV = 0$ vanishes and the remaining two resonances merge into a single peak when T exceeds the Majorana overlap energy ($T > \varepsilon_M$), as was reported before in [61] for weak $|\lambda|$ couplings. When the EPI is switched on ($\beta \neq 0$), we see in Fig. 2.25a for unhybridized MBSs at magnetic flux phase $\phi = 0$ that the two main MBS-associated peaks are more narrow compared to the EPI absence case. Beside the sharp resonances, two large broadened peaks show up in the spectrum of dI/dV at voltages $eV \approx \pm 2|\tilde{\varepsilon}_d|$ owing to the EPI. Furthermore, a series of phonon-assisted additional transport channels develop in the differential conductance spectrum at bias voltages $eV \approx \pm 2(|\tilde{\varepsilon}_d| + p\,\omega_0)$ (where p is an integer). These transport channels correspond to the tunneling of electrons through the dot by emission or absorption of phonons at frequency ω_0. In the vicinity of side-band peaks, characteristic MZM-induced resonances emerge at biases $eV \approx \pm 2|p|\omega_0$. As we have discussed above regarding the linear conductance $\mathcal{G}$, the zero-bias differential conductance at zero temperature is not influenced by the EPI. In the case of hybridized Majoranas [see Fig. 2.25b], the central three-peak structure, appearing for $\beta = 0$, transforms into a spectrum of two sharp MZM-associated resonances with broadened side-peaks at voltages $eV \approx \pm 2|\tilde{\varepsilon}_d|$ because of the presence of EPI. As the temperature increases, these two sharp resonances merge into one. The differential conductance peaks are completely smeared at temperatures higher than the Majorana overlap energy, i.e., $T > \varepsilon_M$. The phonon-assisted satellite peaks follow the shape of main peak. At $\phi = \pi$ for unhybridized MBSs [see Fig. 2.25c], the main two-peak structure developing in the spectrum of differential conductance without EPI, evolves into a peak-structure that has a sharp resonance at the zero bias $eV = 0$ and also two side-resonances at voltages $eV \approx \pm 2|\tilde{\varepsilon}_d|$. One observes that the zero-bias differential conductance shows immunity to the presence of EPI at zero temperature. Similarly, the phonon side-bands emerging in the spectrum of dI/dV are characterized by the central MZM-induced resonances. For overlapping Majoranas [see Fig. 2.25d], we observe that the main three-peak structure, appearing for $\beta = 0$, transforms into a four-peak one in the presence of EPI, which consists of two narrow MZM resonances and two broadened peaks. The phonon-assisted satellite peaks reproduce the main peak shape. The resonances are totally washed out at higher temperatures.

2.4 Summary

In this chapter, we have theoretically analyzed the transport properties of a Majorana circuit formed by two MZMs, located at the opposite sites of a TSNW, connected laterally to a QD in the presence of EPI. The QD-TSNW junction forms a ring structure that is threaded by a tunable magnetic flux, which allows us to control the transport properties of the system. The QD electron interacts with a single long-wave optical phonon mode, leading to phonon-assisted transport processes. We have shown that for an oppositely biased QD system, the tunneling current is a result of two processes, the ET and LAR, in the subgap regime. An asymmetrically biasing of the dot leads to finite contributions of the CAR processes to the tunneling current. In particular, we have turned our attention to the study of transport properties in a system where the QD connects to one of the ends of a TSNW in the absence and presence of electron-phonon coupling. The EPI is incorporated in the model employing a canonical transformation with the nonequilibrium Green's function formalism. The corresponding QD retarded Green's functions have been calculated by applying the EOM technique. The effect of EPI on transport characteristics of the system in the presence of MBSs has been studied in detail. For comparison, we have also investigated a system both with and without EPI.

In the absence of EPI, the normalized spectral function of the QD displays a typical lineshape characterizing MBSs. The position of antiresonance points that develop in the normalized spectral function is determined by the Majorana overlap energy. We have obtained that when the QD couples to both MBSs the linear conductance at zero temperature is in equal manner attributed to ET and LAR processes for unhybridized Majoranas. The linear conductance with its components presents a 2π periodicity as a function of magnetic flux phase, and does not depend on finite values of QD-MBS couplings or QD energy at zero temperature. It has also been found that the ET and LAR linear conductances respond differently to the changes in temperature. Specifically, the ET conductance exceeds the $e^2/4h$ value with the increase of temperature, while the LAR does not, when the dot weakly connects to both MBSs. Note that at finite temperature, the resonances of linear conductance broaden with increasing temperature, and depend on the QD energy and finite values of the dot-MZM coupling strengths even for unhybridized Majoranas. The 2π periodicity of linear conductance becomes 4π when the dot energy is nonzero for hybridized MBSs. We have found that the differential conductance presents a 2π periodicity for nonoverlapping MZMs independent of the value of QD energy. The 2π periodicity transforms to 4π when the QD level is nonzero for overlapping MZMs. The differential conductance resonances attributed to the overlapping Majoranas are smeared when the system temperature surpasses the value of overlap energy, as it was found in the literature [61]. Note that the periodicity of differential and linear conductances remains unchanged for an asymmetrically attached QD-leads junction which would facilitate the experimental verification. In particular, when the dot couples to one of the sites of a TSNW, we found that the dot-MBS coupling suppresses the magnitude of the current, when the QD level is located near the Fermi energy, especially at low biases. This suppression effect can be counteracted by the increase of bias voltage. We have observed that the dot-MZM coupling and Majorana overlap energy between MBS wavefunctions influence significantly the transport properties in the low bias regime.

In the presence of EPI, the normalized spectral function of the QD is visibly changed. Beside the resonances associated to the MBSs, a series of additional peaks, corresponding to phonon-assisted transport channels, emerge in the spectrum. The shape of these peaks is influenced by the values of EPI coupling strength, temperature, Majorana overlap energy, and QD energy level. Most interestingly, when the QD couples to both MBSs, the zero-temperature linear conductance exhibits a 2π periodicity in magnetic flux phase, and is immune to the presence of EPI, changes in QD energy level, and finite values of dot-MZM couplings for unhybridized MZMs. In contrast, at finite temperatures, the linear conductance depends on electron-phonon coupling strength and QD energy for nonoverlapping MBSs. In the case of overlapping Majoranas, the linear conductance presents a 4π periodicity in magnetic flux phase even for $\varepsilon_d = 0$, because the QD energy is renormalized by the electron-phonon coupling, at zero and nonzero temperatures. As a consequence, the periodicity of the linear conductance may transform to 4π for hybridized MBSs when changing the electron-phonon coupling strength or gate voltage. Furthermore, a transition in periodicity of differential conductance between 2π and π for degenerate MBSs, as well as, between 2π and 4π for overlapping MBSs, can be realized by varying the electron-phonon coupling strength. The differential conductance presents phonon-assisted satellite resonances due to the EPI, whose height is also sensitive to the change in temperature and strength of EPI. The magnitude of phonon-assisted peaks in differential conductance spectrum decreases with increasing temperature, and is suppressed at high temperatures. In the case when the QD connects to one of the edges of a TSNW, the current–bias voltage characteristics show a step-like structure under EPI corresponding to phonon-assisted tunneling through a QD. The influence of QD-MBS coupling and Majorana overlap energy on the current versus bias voltage characteristics changes when varying the bias. In addition, the current is not sensitive to the charge fluctuations in the high-voltage regime. We have established that the current reveals a nonmonotonic behavior with the variation of temperature depending on the dot-MZM coupling, Majorana overlap energy, QD level and bias voltage values. Additionally, the current versus temperature characteristics are strongly influenced even by the lower QD energy variations.

The device geometries studied in this chapter, where QDs are integrated in semiconducting-superconducting heterostructures, should be experimentally feasible by taking into consideration the current technology advancements in the field [114, 121, 123, 125, 190]. The influence of an environmental phonon bath on the physical properties of such devices could thus be considered within theoretical models such as those illustrated in Fig. 2.1. These findings presented in this chapter provide more clarity on the expected behavior of MBS fingerprints in transport experiments where EPI plays a central role. Exploring the parameter regimes of such QD-MBS junctions that enable a categorical demonstration of the presence or absence of Majoranas in the system through transport measurements is a promising direction for future research.

Appendix

As mentioned in Sect. 2.2.3, the EOM method allows us to analytically determine the single-particle Green's function for a quantum system governed by the Hamiltonian $\mathcal{H}$. We first define the retarded Green's function for fermionic operators A and B as

$$\langle\langle A(t)|B(t')\rangle\rangle_t^r = -\frac{i}{\hbar}\theta(t-t')\langle\{A(t), B(t')\}\rangle, \tag{2.34}$$

where $\theta(t-t')$ and $\{A, B\} = AB + BA = \{B, A\}$ represent the Heaviside step function and the anticommutator of operators A and B, respectively [159, 181, 182, 191]. Taking the time derivative of $\langle\langle A(t)|B(t')\rangle\rangle_t^r$, leads to the following EOM

$$i\hbar\frac{\partial}{\partial t}\langle\langle A(t)|B(t')\rangle\rangle_t^r = \delta(t-t')\langle\{A, B\}\rangle + \langle\langle[A(t), \mathcal{H}]|B(t')\rangle\rangle_t^r, \tag{2.35}$$

where $[A(t), \mathcal{H}] = A(t)\mathcal{H} - \mathcal{H}A(t) = -[\mathcal{H}, A(t)]$ denotes the commutator of $A(t)$ and $\mathcal{H}$. The Fourier transform of relation (2.35) gives the EOM in energy space

$$\varepsilon^+\langle\langle A|B\rangle\rangle_\varepsilon^r + \langle\langle[\mathcal{H}, A]|B\rangle\rangle_\varepsilon^r = \langle\{A, B\}\rangle, \tag{2.36}$$

where $\varepsilon^+ = \varepsilon + i\delta$, with δ being a positive infinitesimal and $\langle\langle A|B\rangle\rangle_\varepsilon^r$ denotes the Fourier transform of $\langle\langle A(t)|B(t')\rangle\rangle_t^r$. The commutator and anticommutator of operators A, B and C satisfy the following fundamental properties:

$$\begin{aligned}
[A+B, C] &= [A, C] + [B, C], \\
[AB, C] &= A[B, C] + [A, C]B, \\
[A, BC] &= [A, B]C + B[A, C] \\
&= \{A, B\}C - B\{A, C\}, \\
\{A, BC\} &= \{A, B\}C - B[A, C],
\end{aligned} \tag{2.37}$$

which will be used in the following.

In this appendix, based on our paper [159], we calculate the retarded Green's functions of the dot, given in Eqs. (2.29a)–(2.29c), for arbitrary values of ε_M, λ_1, and λ_2 using the EOM method discussed above. To do this, we define the electron-electron and electron-hole retarded Green's functions for the dot as $\tilde{G}_{d11}^r(\varepsilon) = \langle\langle d|d^\dagger\rangle\rangle_\varepsilon^r$ and $\tilde{G}_{d12}^r(\varepsilon) = \langle\langle d|d\rangle\rangle_\varepsilon^r$ by replacing $A(B)$ in $\langle\langle A|B\rangle\rangle_\varepsilon^r$ with $d(d^\dagger)$ and $d(d)$, respectively. The EOM for $\tilde{G}_{d11}^r(\varepsilon)$ is expressed as

$$\varepsilon^+\langle\langle d|d^\dagger\rangle\rangle_\varepsilon^r + \langle\langle[\bar{\mathcal{H}}_{\mathrm{el}}, d\,]|d^\dagger\rangle\rangle_\varepsilon^r = \langle\{d, d^\dagger\}\rangle, \tag{2.38}$$

where $\{d, d^\dagger\} = 1$ and using Eqs. (2.37) and (2.21), one obtains

$$[\bar{\mathcal{H}}_{\rm el}, d\,] = [\mathcal{H}_{\rm Leads}, d\,] + [\mathcal{H}_{\rm MBS}, d\,] + [\tilde{\mathcal{H}}_{\rm QD}, d\,] + [\bar{\mathcal{H}}_{\rm T}, d\,], \tag{2.39}$$

with

$$\begin{aligned}
[\mathcal{H}_{\rm Leads}, d\,] &= \sum_{\alpha,k} \varepsilon_{\alpha k}\big[c^\dagger_{\alpha k}c_{\alpha k}, d\,\big] = -\sum_{\alpha,k} \varepsilon_{\alpha k}\big(\underbrace{\{d, c^\dagger_{\alpha k}\}}_{=0}c_{\alpha k} - c^\dagger_{\alpha k}\underbrace{\{d, c_{\alpha k}\}}_{=0}\big) = 0,\\
[\mathcal{H}_{\rm MBS}, d\,] &= \Big[\varepsilon_M\Big(f^\dagger f - \frac{1}{2}\Big), d\Big] = -\varepsilon_M\big(\underbrace{\{d, f^\dagger\}}_{=0} f - f^\dagger \underbrace{\{d, f\}}_{=0}\big) = 0,\\
[\tilde{\mathcal{H}}_{\rm QD}, d\,] &= \tilde{\varepsilon}_d\big[d^\dagger d, d\,\big] = -\tilde{\varepsilon}_d\big(\underbrace{\{d, d^\dagger\}}_{=1} d - d^\dagger \underbrace{\{d, d\}}_{=0}\big) = -\tilde{\varepsilon}_d\, d,
\end{aligned} \tag{2.40}$$

and

$$\begin{aligned}
[\bar{\mathcal{H}}_{\rm T}, d\,] =& \frac{(\tilde{\lambda}_1 + \tilde{\lambda}_2)}{\sqrt{2}}[\,d\,f, d\,] + \frac{(\tilde{\lambda}_1 - \tilde{\lambda}_2)}{\sqrt{2}}[\,d\,f^\dagger, d\,] - \frac{(\tilde{\lambda}^*_1 - \tilde{\lambda}^*_2)}{\sqrt{2}}[\,d^\dagger f, d\,]\\
& - \frac{(\tilde{\lambda}^*_1 + \tilde{\lambda}^*_2)}{\sqrt{2}}[\,d^\dagger f^\dagger, d\,] + \sum_{\alpha,k}\big(\tilde{V}_{\alpha k}\big[c^\dagger_{\alpha k}d, d\,\big] + \tilde{V}^*_{\alpha k}\big[d^\dagger c_{\alpha k}, d\,\big]\big)\\
=& -\frac{(\tilde{\lambda}_1 + \tilde{\lambda}_2)}{\sqrt{2}}\big(\underbrace{\{d, d\}}_{=0} f - d\underbrace{\{d, f\}}_{=0}\big) - \frac{(\tilde{\lambda}_1 - \tilde{\lambda}_2)}{\sqrt{2}}\big(\underbrace{\{d, d\}}_{=0} f^\dagger - d\underbrace{\{d, f^\dagger\}}_{=0}\big)\\
& + \frac{(\tilde{\lambda}^*_1 - \tilde{\lambda}^*_2)}{\sqrt{2}}\big(\underbrace{\{d, d^\dagger\}}_{=1} f - d^\dagger\underbrace{\{d, f\}}_{=0}\big) + \frac{(\tilde{\lambda}^*_1 + \tilde{\lambda}^*_2)}{\sqrt{2}}\big(\underbrace{\{d, d^\dagger\}}_{=1} f^\dagger - d^\dagger\underbrace{\{d, f^\dagger\}}_{=0}\big)\\
& - \sum_{\alpha,k}\Big[\tilde{V}_{\alpha k}\big(\underbrace{\{d, c^\dagger_{\alpha k}\}}_{=0} d - c^\dagger_{\alpha k}\underbrace{\{d, d\}}_{=0}\big) + \tilde{V}^*_{\alpha k}\big(\underbrace{\{d, d^\dagger\}}_{=1} c_{\alpha k} - d^\dagger\underbrace{\{d, c_{\alpha k}\}}_{=0}\big)\Big]\\
=& \frac{(\tilde{\lambda}^*_1 - \tilde{\lambda}^*_2)}{\sqrt{2}} f + \frac{(\tilde{\lambda}^*_1 + \tilde{\lambda}^*_2)}{\sqrt{2}} f^\dagger - \sum_{\alpha,k}\tilde{V}^*_{\alpha k}c_{\alpha k}.
\end{aligned} \tag{2.41}$$

Substituting relations from Eqs. (2.40) and (2.41) into Eqs. (2.39), (2.38) reads

$$\begin{aligned}
(\varepsilon^+ - \tilde{\varepsilon}_d)\langle\langle d|d^\dagger\rangle\rangle^r_\varepsilon - \sum_{\alpha,k}\tilde{V}^*_{\alpha k}\langle\langle c_{\alpha k}|d^\dagger\rangle\rangle^r_\varepsilon + \frac{1}{\sqrt{2}}(\tilde{\lambda}^*_1 - \tilde{\lambda}^*_2)\langle\langle f|d^\dagger\rangle\rangle^r_\varepsilon\\
+ \frac{1}{\sqrt{2}}(\tilde{\lambda}^*_1 + \tilde{\lambda}^*_2)\langle\langle f^\dagger|d^\dagger\rangle\rangle^r_\varepsilon = 1.
\end{aligned} \tag{2.42}$$

To determine $\langle\langle d|d^\dagger\rangle\rangle^r_\varepsilon$, we need to calculate the new higher-order correlation functions $\langle\langle c_{\alpha k}|d^\dagger\rangle\rangle^r_\varepsilon$, $\langle\langle f|d^\dagger\rangle\rangle^r_\varepsilon$ and $\langle\langle f^\dagger|d^\dagger\rangle\rangle^r_\varepsilon$, appearing in Eq. (2.42). We first write down an equation for the term $\langle\langle c_{\alpha k}|d^\dagger\rangle\rangle^r_\varepsilon$ as

$$\varepsilon^+\langle\langle c_{\alpha k}|d^\dagger\rangle\rangle^r_\varepsilon + \langle\langle[\bar{\mathcal{H}}_{\rm el}, c_{\alpha k}\,]|d^\dagger\rangle\rangle^r_\varepsilon = \langle\{c_{\alpha k}, d^\dagger\}\rangle, \tag{2.43}$$

where $\{c_{\alpha k}, d^\dagger\} = 0$ and

$$[\bar{\mathcal{H}}_{\rm el}, c_{\alpha k}] = [\mathcal{H}_{\rm Leads}, c_{\alpha k}] + [\mathcal{H}_{\rm MBS}, c_{\alpha k}] + [\tilde{\mathcal{H}}_{\rm QD}, c_{\alpha k}] + [\bar{\mathcal{H}}_{\rm T}, c_{\alpha k}], \quad (2.44)$$

with

$$\begin{aligned}
[\mathcal{H}_{\rm Leads}, c_{\alpha k}] &= \sum_{\beta,q} \varepsilon_{\beta q}[c^\dagger_{\beta q} c_{\beta q}, c_{\alpha k}] = -\sum_{\beta,q} \varepsilon_{\beta q}\big(\underbrace{\{c_{\alpha k}, c^\dagger_{\beta q}\}}_{=\delta_{\alpha\beta}\,\delta_{kq}} c_{\beta q} - c^\dagger_{\beta q}\underbrace{\{c_{\alpha k}, c_{\beta q}\}}_{=0}\big) = \\
&= -\varepsilon_{\alpha k}\, c_{\alpha k}, \\
[\mathcal{H}_{\rm MBS}, c_{\alpha k}] &= \Big[\varepsilon_M\Big(f^\dagger f - \frac{1}{2}\Big), c_{\alpha k}\Big] = -\varepsilon_M\big(\underbrace{\{c_{\alpha k}, f^\dagger\}}_{=0} f - f^\dagger \underbrace{\{c_{\alpha k}, f\}}_{=0}\big) = 0, \\
[\tilde{\mathcal{H}}_{\rm QD}, c_{\alpha k}] &= \tilde{\varepsilon}_d[d^\dagger d, c_{\alpha k}] = -\tilde{\varepsilon}_d\big(\underbrace{\{c_{\alpha k}, d^\dagger\}}_{=0} d - d^\dagger \underbrace{\{c_{\alpha k}, d\}}_{=0}\big) = 0,
\end{aligned} \quad (2.45)$$

and

$$\begin{aligned}
[\bar{\mathcal{H}}_{\rm T}, c_{\alpha k}] &= \frac{(\tilde{\lambda}_1 + \tilde{\lambda}_2)}{\sqrt{2}}[d\, f, c_{\alpha k}] + \frac{(\tilde{\lambda}_1 - \tilde{\lambda}_2)}{\sqrt{2}}[d\, f^\dagger, c_{\alpha k}] - \frac{(\tilde{\lambda}_1^* - \tilde{\lambda}_2^*)}{\sqrt{2}}[d^\dagger f, c_{\alpha k}] \\
&- \frac{(\tilde{\lambda}_1^* + \tilde{\lambda}_2^*)}{\sqrt{2}}[d^\dagger f^\dagger, c_{\alpha k}] + \sum_{\beta,q}(\tilde{V}_{\beta q}[c^\dagger_{\beta q} d, c_{\alpha k}] + \tilde{V}^*_{\beta q}[d^\dagger c_{\beta q}, c_{\alpha k}]) \\
&= -\frac{(\tilde{\lambda}_1 + \tilde{\lambda}_2)}{\sqrt{2}}\big(\underbrace{\{c_{\alpha k}, d\}}_{=0} f - d \underbrace{\{c_{\alpha k}, f\}}_{=0}\big) - \frac{(\tilde{\lambda}_1 - \tilde{\lambda}_2)}{\sqrt{2}}\big(\underbrace{\{c_{\alpha k}, d\}}_{=0} f^\dagger - d \underbrace{\{c_{\alpha k}, f^\dagger\}}_{=0}\big) \\
&+ \frac{(\tilde{\lambda}_1^* - \tilde{\lambda}_2^*)}{\sqrt{2}}\big(\underbrace{\{c_{\alpha k}, d^\dagger\}}_{=0} f - d^\dagger \underbrace{\{c_{\alpha k}, f\}}_{=0}\big) + \frac{(\tilde{\lambda}_1^* + \tilde{\lambda}_2^*)}{\sqrt{2}}\big(\underbrace{\{c_{\alpha k}, d^\dagger\}}_{=0} f^\dagger - d^\dagger \underbrace{\{c_{\alpha k}, f^\dagger\}}_{=0}\big) \\
&- \sum_{\beta,q}\Big[\tilde{V}_{\beta q}\big(\underbrace{\{c_{\alpha k}, c^\dagger_{\beta q}\}}_{=\delta_{\alpha\beta}\,\delta_{kq}} d - c^\dagger_{\beta q}\underbrace{\{c_{\alpha k}, d\}}_{=0}\big) + \tilde{V}^*_{\beta q}\big(\underbrace{\{c_{\alpha k}, d^\dagger\}}_{=0} c_{\beta q} - d^\dagger \underbrace{\{c_{\alpha k}, c_{\beta q}\}}_{=0}\big)\Big] \\
&= -\tilde{V}_{\alpha k} d.
\end{aligned} \quad (2.46)$$

Substituting relations from Eqs. (2.45) and (2.46) into Eqs. (2.44), (2.43) becomes

$$\langle\langle c_{\alpha k}|d^\dagger\rangle\rangle^r_\varepsilon = \frac{\tilde{V}_{\alpha k}}{\varepsilon^+ - \varepsilon_{\alpha k}}\langle\langle d|d^\dagger\rangle\rangle^r_\varepsilon. \quad (2.47)$$

Using Eq. (2.47) with Eq. (2.42), in the wide-band limit, we can write [166]

$$\sum_{\alpha,k} \tilde{V}^*_{\alpha k}\langle\langle c_{\alpha k}|d^\dagger\rangle\rangle^r_\varepsilon \approx -i\tilde{\Gamma}\langle\langle d|d^\dagger\rangle\rangle^r_\varepsilon, \quad (2.48)$$

where $\tilde{\Gamma} = \frac{1}{2}\sum_\alpha \tilde{\Gamma}_\alpha$ with $\tilde{\Gamma}_\alpha = 2\pi\sum_k |\tilde{V}_{\alpha k}|^2\delta(\varepsilon - \varepsilon_{\alpha k})$ being the coupling between the QD and lead α, where we assumed electron-hole symmetry in the dot-lead couplings ($\Gamma_\alpha = \Gamma_\alpha^e = \Gamma_\alpha^h$), as mentioned in Sect. 2.2.2. Substituting now Eq. (2.48) into Eq. (2.42), one finds

$$(\varepsilon^+ - \tilde{\varepsilon}_d + i\tilde{\Gamma})\langle\langle d|d^\dagger\rangle\rangle^r_\varepsilon + \frac{1}{\sqrt{2}}(\tilde{\lambda}_1^* - \tilde{\lambda}_2^*)\langle\langle f|d^\dagger\rangle\rangle^r_\varepsilon + \frac{1}{\sqrt{2}}(\tilde{\lambda}_1^* + \tilde{\lambda}_2^*)\langle\langle f^\dagger|d^\dagger\rangle\rangle^r_\varepsilon = 1. \tag{2.49}$$

To obtain an analytical formula for $\langle\langle d|d^\dagger\rangle\rangle^r_\varepsilon$, we need to calculate the remaining terms, $\langle\langle f|d^\dagger\rangle\rangle^r_\varepsilon$ and $\langle\langle f^\dagger|d^\dagger\rangle\rangle^r_\varepsilon$, in Eq. (2.49). The equation for $\langle\langle f|d^\dagger\rangle\rangle^r_\varepsilon$ reads

$$\varepsilon^+\langle\langle f|d^\dagger\rangle\rangle^r_\varepsilon + \langle\langle[\bar{\mathcal{H}}_{\text{el}}, f]|d^\dagger\rangle\rangle^r_\varepsilon = \langle\{f, d^\dagger\}\rangle, \tag{2.50}$$

where $\{f, d^\dagger\} = 0$ and

$$[\bar{\mathcal{H}}_{\text{el}}, f] = [\mathcal{H}_{\text{Leads}}, f] + [\mathcal{H}_{\text{MBS}}, f] + [\tilde{\mathcal{H}}_{\text{QD}}, f] + [\bar{\mathcal{H}}_{\text{T}}, f], \tag{2.51}$$

with

$$\begin{aligned}
[\mathcal{H}_{\text{Leads}}, f] &= \sum_{\alpha,k} \varepsilon_{\alpha k}[c^\dagger_{\alpha k}c_{\alpha k}, f] = -\sum_{\alpha,k}\varepsilon_{\alpha k}\big(\underbrace{\{f, c^\dagger_{\alpha k}\}}_{=0}c_{\alpha k} - c^\dagger_{\alpha k}\underbrace{\{f, c_{\alpha k}\}}_{=0}\big) = 0,\\
[\mathcal{H}_{\text{MBS}}, f] &= \Big[\varepsilon_M\Big(f^\dagger f - \frac{1}{2}\Big), f\Big] = -\varepsilon_M\big(\underbrace{\{f, f^\dagger\}}_{=1}f - f^\dagger\underbrace{\{f, f\}}_{=0}\big) = -\varepsilon_M f,\\
[\tilde{\mathcal{H}}_{\text{QD}}, f] &= \tilde{\varepsilon}_d[d^\dagger d, f] = -\tilde{\varepsilon}_d\big(\underbrace{\{f, d^\dagger\}}_{=0}d - d^\dagger\underbrace{\{f, d\}}_{=0}\big) = 0,
\end{aligned} \tag{2.52}$$

and

$$\begin{aligned}
[\bar{\mathcal{H}}_{\text{T}}, f] &= \frac{(\tilde{\lambda}_1 + \tilde{\lambda}_2)}{\sqrt{2}}[d\, f, f] + \frac{(\tilde{\lambda}_1 - \tilde{\lambda}_2)}{\sqrt{2}}[d\, f^\dagger, f] - \frac{(\tilde{\lambda}_1^* - \tilde{\lambda}_2^*)}{\sqrt{2}}[d^\dagger f, f]\\
&\quad - \frac{(\tilde{\lambda}_1^* + \tilde{\lambda}_2^*)}{\sqrt{2}}[d^\dagger f^\dagger, f] + \sum_{\alpha,k}\big(\tilde{V}_{\alpha k}[c^\dagger_{\alpha k}d, f] + \tilde{V}^*_{\alpha k}[d^\dagger c_{\alpha k}, f]\big)\\
&= -\frac{(\tilde{\lambda}_1 + \tilde{\lambda}_2)}{\sqrt{2}}\big(\underbrace{\{f, d\}}_{=0}f - d\underbrace{\{f, f\}}_{=0}\big) - \frac{(\tilde{\lambda}_1 - \tilde{\lambda}_2)}{\sqrt{2}}\big(\underbrace{\{f, d\}}_{=0}f^\dagger - d\underbrace{\{f, f^\dagger\}}_{=1}\big)\\
&\quad + \frac{(\tilde{\lambda}_1^* - \tilde{\lambda}_2^*)}{\sqrt{2}}\big(\underbrace{\{f, d^\dagger\}}_{=0}f - d^\dagger\underbrace{\{f, f\}}_{=0}\big) + \frac{(\tilde{\lambda}_1^* + \tilde{\lambda}_2^*)}{\sqrt{2}}\big(\underbrace{\{f, d^\dagger\}}_{=0}f^\dagger - d^\dagger\underbrace{\{f, f^\dagger\}}_{=1}\big)\\
&\quad - \sum_{\alpha,k}\Big[\tilde{V}_{\alpha k}\big(\underbrace{\{f, c^\dagger_{\alpha k}\}}_{=0}d - c^\dagger_{\alpha k}\underbrace{\{f, d\}}_{=0}\big) + \tilde{V}^*_{\alpha k}\big(\underbrace{\{f, d^\dagger\}}_{=0}c_{\alpha k} - d^\dagger\underbrace{\{f, c_{\alpha k}\}}_{=0}\big)\Big]\\
&= \frac{(\tilde{\lambda}_1 - \tilde{\lambda}_2)}{\sqrt{2}}d - \frac{(\tilde{\lambda}_1^* + \tilde{\lambda}_2^*)}{\sqrt{2}}d^\dagger.
\end{aligned} \tag{2.53}$$

Substituting relations from Eqs. (2.52) and (2.53) into Eqs. (2.51), (2.50) reads

$$(\varepsilon^+ - \varepsilon_M)\langle\langle f|d^\dagger\rangle\rangle^r_\varepsilon - \frac{1}{\sqrt{2}}(\tilde{\lambda}_1^* + \tilde{\lambda}_2^*)\langle\langle d^\dagger|d^\dagger\rangle\rangle^r_\varepsilon + \frac{1}{\sqrt{2}}(\tilde{\lambda}_1 - \tilde{\lambda}_2)\langle\langle d|d^\dagger\rangle\rangle^r_\varepsilon = 0. \tag{2.54}$$

Similarly to the derivation of the equation for $\langle\langle f|d^\dagger\rangle\rangle^r_\varepsilon$, using the relations $\{f^\dagger, d^\dagger\} = 0$, $\{f^\dagger, c^\dagger_{\alpha k}\} = 0$ and $\{f^\dagger, f^\dagger\} = 0$, the EOM for $\langle\langle f^\dagger|d^\dagger\rangle\rangle^r_\varepsilon$ is expressed as

$$(\varepsilon^+ + \varepsilon_M)\langle\langle f^\dagger|d^\dagger\rangle\rangle^r_\varepsilon - \frac{1}{\sqrt{2}}(\tilde{\lambda}_1^* - \tilde{\lambda}_2^*)\langle\langle d^\dagger|d^\dagger\rangle\rangle^r_\varepsilon + \frac{1}{\sqrt{2}}(\tilde{\lambda}_1 + \tilde{\lambda}_2)\langle\langle d|d^\dagger\rangle\rangle^r_\varepsilon = 0. \tag{2.55}$$

Now, similarly to Eq. (2.42), we can write an equation for the correlation function $\langle\langle d^\dagger|d^\dagger\rangle\rangle^r_\varepsilon$ appearing in Eqs. (2.54) and (2.55):

$$\begin{aligned}(\varepsilon^+ + \tilde{\varepsilon}_d)\langle\langle d^\dagger|d^\dagger\rangle\rangle^r_\varepsilon + \sum_{\alpha,k}\tilde{V}_{\alpha k}\langle\langle c^\dagger_{\alpha k}|d^\dagger\rangle\rangle^r_\varepsilon - \frac{1}{\sqrt{2}}(\tilde{\lambda}_1 + \tilde{\lambda}_2)\langle\langle f|d^\dagger\rangle\rangle^r_\varepsilon \\ - \frac{1}{\sqrt{2}}(\tilde{\lambda}_1 - \tilde{\lambda}_2)\langle\langle f^\dagger|d^\dagger\rangle\rangle^r_\varepsilon = 0.\end{aligned} \tag{2.56}$$

Similarly to $\langle\langle c_{\alpha k}|d^\dagger\rangle\rangle^r_\varepsilon$, the EOM for the term $\langle\langle c^\dagger_{\alpha k}|d^\dagger\rangle\rangle^r_\varepsilon$ in Eq. (2.56) is written as

$$\langle\langle c^\dagger_{\alpha k}|d^\dagger\rangle\rangle^r_\varepsilon = -\frac{\tilde{V}^*_{\alpha k}}{\varepsilon^+ + \varepsilon_{\alpha k}}\langle\langle d^\dagger|d^\dagger\rangle\rangle^r_\varepsilon. \tag{2.57}$$

Here, assuming again electron-hole symmetry in QD-lead couplings and substituting Eq. (2.57) into Eq. (2.56), one obtains

$$(\varepsilon^+ + \tilde{\varepsilon}_d + i\tilde{\Gamma})\langle\langle d^\dagger|d^\dagger\rangle\rangle^r_\varepsilon - \frac{1}{\sqrt{2}}(\tilde{\lambda}_1 + \tilde{\lambda}_2)\langle\langle f|d^\dagger\rangle\rangle^r_\varepsilon - \frac{1}{\sqrt{2}}(\tilde{\lambda}_1 - \tilde{\lambda}_2)\langle\langle f^\dagger|d^\dagger\rangle\rangle^r_\varepsilon = 0. \tag{2.58}$$

Now, we have a complete set of equations based on the higher-order correlation functions, which leads to expressing an analytical formula for the retarded Green's function $\langle\langle d|d^\dagger\rangle\rangle^r_\varepsilon$. To do this, we substitute $\langle\langle f|d^\dagger\rangle\rangle^r_\varepsilon$ from Eq. (2.54) and $\langle\langle f^\dagger|d^\dagger\rangle\rangle^r_\varepsilon$ from Eq. (2.55) into Eqs. (2.49) and (2.58), and find

$$\begin{aligned}\Big\{(\varepsilon^+ - \tilde{\varepsilon}_d + i\tilde{\Gamma}) - \Big[(|\tilde{\lambda}_1|^2 + |\tilde{\lambda}_2|^2) - \frac{\varepsilon_M}{\varepsilon}(\tilde{\lambda}_1\tilde{\lambda}_2^* + \tilde{\lambda}_1^*\tilde{\lambda}_2)\Big]\mathcal{K}(\varepsilon)\Big\}\langle\langle d|d^\dagger\rangle\rangle^r_\varepsilon \\ + (\tilde{\lambda}_1^*\tilde{\lambda}_1^* - \tilde{\lambda}_2^*\tilde{\lambda}_2^*)\mathcal{K}(\varepsilon)\langle\langle d^\dagger|d^\dagger\rangle\rangle^r_\varepsilon = 1,\end{aligned} \tag{2.59}$$

$$\begin{aligned}\Big\{(\varepsilon^+ + \tilde{\varepsilon}_d + i\tilde{\Gamma}) - \Big[(|\tilde{\lambda}_1|^2 + |\tilde{\lambda}_2|^2) + \frac{\varepsilon_M}{\varepsilon}(\tilde{\lambda}_1\tilde{\lambda}_2^* + \tilde{\lambda}_1^*\tilde{\lambda}_2)\Big]\mathcal{K}(\varepsilon)\Big\}\langle\langle d^\dagger|d^\dagger\rangle\rangle^r_\varepsilon \\ + (\tilde{\lambda}_1\tilde{\lambda}_1 - \tilde{\lambda}_2\tilde{\lambda}_2)\mathcal{K}(\varepsilon)\langle\langle d|d^\dagger\rangle\rangle^r_\varepsilon = 0,\end{aligned} \tag{2.60}$$

with notation:

$$\mathcal{K}(\varepsilon) = \frac{\varepsilon}{\varepsilon^2 - \varepsilon_M^2}. \tag{2.61}$$

Expressing $\langle\langle d^\dagger | d^\dagger \rangle\rangle_\varepsilon^r$ from Eq. (2.60) and replacing it into Eq. (2.59), we obtain for the electron-electron retarded Green's function of the dot

$$\tilde{G}_{d11}^r(\varepsilon) = \left\{(\varepsilon + \tilde{\varepsilon}_d + i\tilde{\Gamma}) - \left[(|\tilde{\lambda}_1|^2 + |\tilde{\lambda}_2|^2) + \frac{\varepsilon_M}{\varepsilon}(\tilde{\lambda}_1\tilde{\lambda}_2^* + \tilde{\lambda}_1^*\tilde{\lambda}_2)\right]\mathcal{K}(\varepsilon)\right\}\tilde{\mathcal{F}}(\varepsilon)^{-1}, \tag{2.62}$$

where

$$\begin{aligned}\tilde{\mathcal{F}}(\varepsilon) &= (\varepsilon - \tilde{\varepsilon}_d + i\tilde{\Gamma})(\varepsilon + \tilde{\varepsilon}_d + i\tilde{\Gamma}) - 2(\varepsilon + i\tilde{\Gamma})(|\tilde{\lambda}_1|^2 + |\tilde{\lambda}_2|^2)\mathcal{K}(\varepsilon) \\ &+ \frac{2\varepsilon_M}{\varepsilon}\tilde{\varepsilon}_d(\tilde{\lambda}_1\tilde{\lambda}_2^* + \tilde{\lambda}_1^*\tilde{\lambda}_2)\mathcal{K}(\varepsilon) + \frac{1}{\varepsilon}(2|\tilde{\lambda}_1|^2|\tilde{\lambda}_2|^2 + \tilde{\lambda}_1\tilde{\lambda}_1\tilde{\lambda}_2^*\tilde{\lambda}_2^* + \tilde{\lambda}_1^*\tilde{\lambda}_1^*\tilde{\lambda}_2\tilde{\lambda}_2)\mathcal{K}(\varepsilon).\end{aligned} \tag{2.63}$$

Following the steps presented above for $\tilde{G}_{d11}^r(\varepsilon) = \langle\langle d | d^\dagger \rangle\rangle_\varepsilon^r$, the EOM for electron-hole retarded Green's function of the dot $\tilde{G}_{d12}^r(\varepsilon) = \langle\langle d | d \rangle\rangle_\varepsilon^r$ reads

$$\begin{aligned}(\varepsilon^+ - \tilde{\varepsilon}_d)\langle\langle d|d\rangle\rangle_\varepsilon^r - \sum_{\alpha,k}\tilde{V}_{\alpha k}^*\langle\langle c_{\alpha k}|d\rangle\rangle_\varepsilon^r + \frac{1}{\sqrt{2}}(\tilde{\lambda}_1^* - \tilde{\lambda}_2^*)\langle\langle f|d\rangle\rangle_\varepsilon^r \\ + \frac{1}{\sqrt{2}}(\tilde{\lambda}_1^* + \tilde{\lambda}_2^*)\langle\langle f^\dagger|d\rangle\rangle_\varepsilon^r = 0,\end{aligned} \tag{2.64}$$

where the equation for term $\langle\langle c_{\alpha k}|d\rangle\rangle_\varepsilon^r$ is

$$\langle\langle c_{\alpha k}|d\rangle\rangle_\varepsilon^r = \frac{\tilde{V}_{\alpha k}}{\varepsilon^+ - \varepsilon_{\alpha k}}\langle\langle d|d\rangle\rangle_\varepsilon^r. \tag{2.65}$$

Substituting Eq. (2.65) into relation (2.64), one finds

$$(\varepsilon^+ - \tilde{\varepsilon}_d + i\tilde{\Gamma})\langle\langle d|d\rangle\rangle_\varepsilon^r + \frac{1}{\sqrt{2}}(\tilde{\lambda}_1^* - \tilde{\lambda}_2^*)\langle\langle f|d\rangle\rangle_\varepsilon^r + \frac{1}{\sqrt{2}}(\tilde{\lambda}_1^* + \tilde{\lambda}_2^*)\langle\langle f^\dagger|d\rangle\rangle_\varepsilon^r = 0. \tag{2.66}$$

The equations for terms $\langle\langle f|d\rangle\rangle_\varepsilon^r$ and $\langle\langle f^\dagger|d\rangle\rangle_\varepsilon^r$ are expressed as

$$(\varepsilon^+ - \varepsilon_M)\langle\langle f|d\rangle\rangle_\varepsilon^r - \frac{1}{\sqrt{2}}(\tilde{\lambda}_1^* + \tilde{\lambda}_2^*)\langle\langle d^\dagger|d\rangle\rangle_\varepsilon^r + \frac{1}{\sqrt{2}}(\tilde{\lambda}_1 - \tilde{\lambda}_2)\langle\langle d|d\rangle\rangle_\varepsilon^r = 0, \tag{2.67}$$

and

$$(\varepsilon^+ + \varepsilon_M)\langle\langle f^\dagger|d\rangle\rangle_\varepsilon^r - \frac{1}{\sqrt{2}}(\tilde{\lambda}_1^* - \tilde{\lambda}_2^*)\langle\langle d^\dagger|d\rangle\rangle_\varepsilon^r + \frac{1}{\sqrt{2}}(\tilde{\lambda}_1 + \tilde{\lambda}_2)\langle\langle d|d\rangle\rangle_\varepsilon^r = 0, \tag{2.68}$$

respectively. The EOM for the term $\langle\langle d^\dagger | d\rangle\rangle^r_\varepsilon$, appearing in Eqs. (2.67) and (2.68), reads

$$\begin{aligned}(\varepsilon^+ + \tilde{\varepsilon}_d)\langle\langle d^\dagger | d\rangle\rangle^r_\varepsilon + \sum_{\alpha,k} \tilde{V}_{\alpha k}\langle\langle c^\dagger_{\alpha k} | d\rangle\rangle^r_\varepsilon - \frac{1}{\sqrt{2}}(\tilde{\lambda}_1 + \tilde{\lambda}_2)\langle\langle f | d\rangle\rangle^r_\varepsilon \\ - \frac{1}{\sqrt{2}}(\tilde{\lambda}_1 - \tilde{\lambda}_2)\langle\langle f^\dagger | d\rangle\rangle^r_\varepsilon = 1,\end{aligned} \tag{2.69}$$

where the equation for the term $\langle\langle c^\dagger_{\alpha k} | d\rangle\rangle^r_\varepsilon$ becomes

$$\langle\langle c^\dagger_{\alpha k} | d\rangle\rangle^r_\varepsilon = -\frac{\tilde{V}^*_{\alpha k}}{\varepsilon^+ + \varepsilon_{\alpha k}}\langle\langle d^\dagger | d\rangle\rangle^r_\varepsilon. \tag{2.70}$$

Substituting relation (2.70) into Eq. (2.69), we find

$$(\varepsilon^+ + \tilde{\varepsilon}_d + i\tilde{\Gamma})\langle\langle d^\dagger | d\rangle\rangle^r_\varepsilon - \frac{1}{\sqrt{2}}(\tilde{\lambda}_1 + \tilde{\lambda}_2)\langle\langle f | d\rangle\rangle^r_\varepsilon - \frac{1}{\sqrt{2}}(\tilde{\lambda}_1 - \tilde{\lambda}_2)\langle\langle f^\dagger | d\rangle\rangle^r_\varepsilon = 1. \tag{2.71}$$

Achieving now a complete set of equations, we can express an analytical formula for the retarded Green's function $\langle\langle d | d\rangle\rangle^r_\varepsilon$. First, we substitute $\langle\langle f | d\rangle\rangle^r_\varepsilon$ from Eq. (2.67) and $\langle\langle f^\dagger | d\rangle\rangle^r_\varepsilon$ from Eq. (2.68) into Eqs. (2.66) and (2.71), and obtain

$$\begin{aligned}\Big\{(\varepsilon^+ - \tilde{\varepsilon}_d + i\tilde{\Gamma}) - \Big[(|\tilde{\lambda}_1|^2 + |\tilde{\lambda}_2|^2) - \frac{\varepsilon_M}{\varepsilon}(\tilde{\lambda}_1\tilde{\lambda}_2^* + \tilde{\lambda}_1^*\tilde{\lambda}_2)\Big]\mathcal{K}(\varepsilon)\Big\}\langle\langle d | d\rangle\rangle^r_\varepsilon \\ + (\tilde{\lambda}_1^*\tilde{\lambda}_1^* - \tilde{\lambda}_2^*\tilde{\lambda}_2^*)\mathcal{K}(\varepsilon)\langle\langle d^\dagger | d\rangle\rangle^r_\varepsilon = 0,\end{aligned} \tag{2.72}$$

and

$$\begin{aligned}\Big\{(\varepsilon^+ + \tilde{\varepsilon}_d + i\tilde{\Gamma}) - \Big[(|\tilde{\lambda}_1|^2 + |\tilde{\lambda}_2|^2) + \frac{\varepsilon_M}{\varepsilon}(\tilde{\lambda}_1\tilde{\lambda}_2^* + \tilde{\lambda}_1^*\tilde{\lambda}_2)\Big]\mathcal{K}(\varepsilon)\Big\}\langle\langle d^\dagger | d\rangle\rangle^r_\varepsilon \\ + (\tilde{\lambda}_1\tilde{\lambda}_1 - \tilde{\lambda}_2\tilde{\lambda}_2)\mathcal{K}(\varepsilon)\langle\langle d | d\rangle\rangle^r_\varepsilon = 1,\end{aligned} \tag{2.73}$$

respectively. Substituting $\langle\langle d^\dagger | d\rangle\rangle^r_\varepsilon$ from Eq. (2.73) into Eq. (2.72) yields the electron-hole retarded Green's function of the dot,

$$\tilde{G}^r_{d12}(\varepsilon) = -(\tilde{\lambda}_1^*\tilde{\lambda}_1^* - \tilde{\lambda}_2^*\tilde{\lambda}_2^*)\mathcal{K}(\varepsilon)\tilde{\mathcal{F}}(\varepsilon)^{-1}. \tag{2.74}$$

References

1. K. Flensberg, Non-Abelian operations on Majorana fermions via single-charge control. Phys. Rev. Lett. **106**, 090503 (2011)
2. D.E. Liu, H.U. Baranger, Detecting a Majorana-fermion zero mode using a quantum dot. Phys. Rev. B **84**, 201308 (2011)
3. M. Leijnse, K. Flensberg, Scheme to measure Majorana fermion lifetimes using a quantum dot. Phys. Rev. B **84**, 140501 (2011)
4. Y. Cao, P. Wang, G. Xiong, M. Gong, X.-Q. Li, Probing the existence and dynamics of Majorana fermion via transport through a quantum dot. Phys. Rev. B **86**, 115311 (2012)
5. M. Lee, J.S. Lim, R. López, Kondo effect in a quantum dot side-coupled to a topological superconductor. Phys. Rev. B **87**, 241402 (2013)
6. M. Cheng, M. Becker, B. Bauer, R.M. Lutchyn, Interplay between Kondo and Majorana interactions in quantum dots. Phys. Rev. X **4**, 031051 (2014)
7. E. Vernek, P.H. Penteado, A.C. Seridonio, J.C. Egues, Subtle leakage of a Majorana mode into a quantum dot. Phys. Rev. B **89**, 165314 (2014)
8. P. Stefański, Signatures of Majorana states in electron transport through a quantum dot coupled to a topological wire. Acta Phys. Pol., A **127**, 198–200 (2015)
9. D. Chevallier, P. Szumniak, S. Hoffman, D. Loss, J. Klinovaja, Topological phase detection in Rashba nanowires with a quantum dot. Phys. Rev. B **97**, 045404 (2018)
10. J.P. Ramos-Andrade, P.A. Orellana, S.E. Ulloa, Detecting coupling of Majorana bound states with an Aharonov–Bohm interferometer. J. Phys.: Condens. Matter. **30**, 045301 (2018)
11. L.S. Ricco, Y. Marques, J.E. Sanches, I.A. Shelykh, A.C. Seridonio, Interaction induced hybridization of Majorana zero modes in a coupled quantum-dot-superconducting-nanowire hybrid system. Phys. Rev. B **102**, 165104 (2020)
12. A. Schuray, M. Rammler, P. Recher, Signatures of the Majorana spin in electrical transport through a Majorana nanowire. Phys. Rev. B **102**, 045303 (2020)
13. L. Liu, X.-F. Chen, J. Song, Y.-X. Li, Signatures of Majorana bound state in an Aharonov-Bohm interferometer embedded with quantum dot. Physica E **124**, 114212 (2020)
14. P. Majek, I. Weymann, Majorana mode leaking into a spin-charge entangled double quantum dot. Phys. Rev. B **104**, 085416 (2021)
15. S. Smirnov, Majorana ensembles with fractional entropy and conductance in nanoscopic systems. Phys. Rev. B **104**, 205406 (2021)
16. T. Gong, L.-L. Zhang, X.-F. Dai, C. Jiang, W.-J. Gong, Double resonant tunneling due to the nonlocal coupling of Majorana bound states to the quantum dot system. Eur. Phys. J. Plus **137**, 122 (2022)
17. T. Gong, L.-L. Zhang, S.-F. Zhang, C. Jiang, W.-J. Gong, Two-stage decoupling effect in electron transport through T-shaped quantum dots coupled via Majorana bound states. Ann. Phys. **447**, 169143 (2022)
18. C. Jiang, G.-Y. Meng, T. Gong, L.-L. Zhang, Zero-bias conductance modification in the quantum-dot system with side-coupled Majorana bound states. Appl. Phys. A **128**, 533 (2022)
19. J.E. Sanches, L.T. Lustosa, L.S. Ricco, I.A. Shelykh, M. de Souza, M.S. Figueira, A.C. Seridonio, Fractionalization of Majorana-Ising-type quasiparticle excitations. Phys. Rev. B **107**, 155144 (2023)
20. A. Huguet, K. Wrześniewski, I. Weymann, Spin effects on transport and zero-bias anomaly in a hybrid Majorana wire-quantum dot system. Sci. Rep. **13**, 17279 (2023)
21. T. Gong, L.-L. Zhang, C. Jiang, S.-F. Zhang, W.-J. Gong, Special odd-even effects of electron transport through quantum-dot molecules with side-coupled Majorana zero modes. Phys. Rev. Res. **5**, 043162 (2023)
22. P. Majek, I. Weymann, Spin-selective transport in a correlated double quantum dot-Majorana wire system. Sci. Rep. **14**, 17762 (2024)
23. Y.-M. Gao, Y.-H. Shen, F. Chi, Z.-C. Yi, L.-M. Liu, Quantum transport through a quantum dot coupled to Majorana nanowire and two ferromagnets with noncollinear magnetization. Nanomaterials **14**(14) (2024)

24. K.P. Wójcik, T. Domański, I. Weymann, Signatures of Kondo-Majorana interplay in ac response. Phys. Rev. B **109**, 075432 (2024)
25. H.-F. Lü, H.-Z. Lu, S.-Q. Shen, Nonlocal noise cross correlation mediated by entangled Majorana fermions. Phys. Rev. B **86**, 075318 (2012)
26. Q. Chen, K.-Q. Chen, H.-K. Zhao, Shot noise in a quantum dot system coupled with Majorana bound states. J. Phys.: Condens. Matter. **26**, 315011 (2014)
27. H.-F. Lü, H.-Z. Lu, S.-Q. Shen, Current noise cross correlation mediated by Majorana bound states. Phys. Rev. B **90**, 195404 (2014)
28. D.E. Liu, M. Cheng, R.M. Lutchyn, Probing Majorana physics in quantum-dot shot-noise experiments. Phys. Rev. B **91**, 081405 (2015)
29. H.-K. Zhao, J. Zhang, J. Wang, Dynamic shot noise in a quantum dot coupled with Majorana fermions under the perturbation of microwave fields. Eur. Phys. J. B **89**, 255 (2016)
30. H.-F. Lü, H.-Z. Lu, S.-Q. Shen, Enhanced current noise correlations in a Coulomb-Majorana device. Phys. Rev. B **93**, 245418 (2016)
31. S. Smirnov, Revealing universal Majorana fractionalization using differential shot noise and conductance in nonequilibrium states controlled by tunneling phases. Phys. Rev. B **105**, 205430 (2022)
32. S. Smirnov, Majorana differential shot noise and its universal thermoelectric crossover. Phys. Rev. B **107**, 155416 (2023)
33. S. Smirnov, Nonequilibrium finite frequency resonances in differential quantum noise driven by Majorana interference. Phys. Rev. B **109**, 195410 (2024)
34. W.-K. Zou, Q. Wang, H.-K. Zhao, Shot noise in Majorana bound state modulated double-quantum-dot interferometer connected to normal and superconductive terminals. Eur. Phys. J. B **97**, 127 (2024)
35. M. Leijnse, Thermoelectric signatures of a Majorana bound state coupled to a quantum dot. New J. Phys. **16**, 015029 (2014)
36. R. López, M. Lee, L. Serra, J.S. Lim, Thermoelectrical detection of Majorana states. Phys. Rev. B **89**, 205418 (2014)
37. H. Khim, R. Lpez, J.S. Lim, M. Lee, Thermoelectric effect in the Kondo dot side-coupled to a Majorana mode. Eur. Phys. J. B **88**(151), 015029 (2015)
38. L.S. Ricco, F.A. Dessotti, I.A. Shelykh, M.S. Figueira, A.C. Seridonio, Tuning of heat and charge transport by Majorana fermions. Sci. Rep. **8**, 2790 (2018)
39. C.W. Ri, K.H. Jong, S.J. Im, H.C. Pak, Thermoelectric transport through a finite-U quantum dot side-coupled to Majorana bound state. Eur. Phys. J. B **92**, 90 (2019)
40. F. Chi, Z.-G. Fu, J. Liu, K.-M. Li, Z. Wang, P. Zhang, Thermoelectric effect in a correlated quantum dot side-coupled to Majorana bound states. Nanoscale Res. Lett. **15**, 79 (2020)
41. P. Majek, K.P. Wójcik, I. Weymann, Spin-resolved thermal signatures of Majorana-Kondo interplay in double quantum dots. Phys. Rev. B **105**, 075418 (2022)
42. I. Grosu, I. Tifrea, Thermoelectric transport properties of single quantum dot systems in the presence of Majorana states. Physica E **145**, 115503 (2023)
43. W.-K. Zou, Q. Wang, H.-K. Zhao, Dynamic heat and charge transports through double-quantum-dot-interferometer modulated by Majorana bound states and time-oscillating Aharonov-Bohm flux. J. Phys.: Condens. Matter **35**, 165303 (2023)
44. C. Wang, X.-Q. Wang, Thermoelectric signature of Majorana zero modes in a T-typed double-quantum-dot structure. Chinese Phys. B **32**, 037304 (2023)
45. P. Trocha, T. Jonckheere, J. Rech, T. Martin, Out-of-equilibrium voltage and thermal bias response of a quantum dot hybrid system coupled to topological superconductor. J. Magn. Magn. Mater. **596**, 171922 (2024)
46. H.-J. Chen, K.-D. Zhu, Nonlinear optomechanical detection for Majorana fermions via a hybrid nanomechanical system. Nanoscale Res. Lett. **9**, 166 (2014)
47. H.-J. Chen, H.-W. Wu, Rabi splitting and optical Kerr nonlinearity of quantum dot mediated by Majorana fermions. Sci. Rep. **8**, 17677 (2018)
48. H.-J. Chen, Majorana fermions induced Fano resonance and fast-to-slow light in a hybrid semiconductor/superconductor ring device. Quantum Inf. Process. **19**, 171 (2020)

49. Y.-L. Lee, Y.-W. Lee, Interplay between Coulomb blockade and Josephson effect in a topological superconductor-quantum dot device. Phys. Rev. B **93**, 184502 (2016)
50. F. Chi, Q.-S. Jia, J. Liu, Q.-G. Gao, Z.-C. Yi, L.-M. Liu, Electronic Tunnelling in Superconductor/Quantum-Dot Josephson Junction Side-Coupled to Majorana Nanowire. Coatings **13**(3), 612 (2023)
51. F. Chi, Q.-S. Jia, J. Liu, Q.-G. Gao, Z.-C. Yi, L.-M. Liu, Enhancement of the Josephson current in a quantum dot connected to Majorana nanowires. Nanomaterials **13**(9) (2023)
52. Y. Gao, X. Zhang, Z. Yi, L. Liu, F. Chi, Thermophase Seebeck coefficient in hybridized superconductor-quantum-dot-superconductor Josephson junction side-coupled to Majorana nanowire. Nanomaterials **13**(17) (2023)
53. G.-Y. Yi, M.-Y. Ma, T. Gong, C. Jiang, W.-J. Gong, Spin-resolved Yu-Shiba-Rusinov states and bistable phase caused by side-coupled Majorana bound states in quantum-dot Josephson junctions. Phys. Rev. B **109**, 165433 (2024)
54. A. Ueda, T. Yokoyama, Anomalous interference in Aharonov-Bohm rings with two Majorana bound states. Phys. Rev. B **90**, 081405 (2014)
55. F.A. Dessotti, L.S. Ricco, M. de Souza, F.M. Souza, A.C. Seridonio, Probing the antisymmetric Fano interference assisted by a Majorana fermion. J. Appl. Phys. **116**(17), 173701 (2014)
56. C. Jiang, Y.-S. Zheng, Fano effect in the Andreev reflection of the Aharonov–Bohm–Fano ring with Majorana bound states. Solid State Commun. **212**, 14–18 (2015)
57. J.-J. Xia, S.-Q. Duan, W. Zhang, Detection of Majorana fermions by Fano resonance in hybrid nanostructures. Nanoscale Res. Lett. **10**(223), 1375 (2015)
58. Q.-B. Zeng, S. Chen, R.L, Fano effect in an AB interferometer with a quantum dot side-coupled to a single Majorana bound state. Phys. Lett. A **380**, 951–957 (2016)
59. Q.-B. Zeng, S. Chen, L. You, R. Lü, Transport through a quantum dot coupled to two Majorana bound states. Front. Phys. **12**(4), 127302 (2017)
60. J. Barański, A. Kobiałka, T. Domański, Spin-sensitive interference due to Majorana state on the interface between normal and superconducting leads. J. Phys.: Condens. Matter. **29**(7), 075603 (2017)
61. A. Schuray, L. Weithofer, P. Recher, Fano resonances in Majorana bound states-quantum dot hybrid systems. Phys. Rev. B **96**, 085417 (2017)
62. L.S. Ricco, V.L. Campo, I.A. Shelykh, A.C. Seridonio, Majorana oscillations modulated by Fano interference and degree of nonlocality in a topological superconducting-nanowire-quantum-dot system. Phys. Rev. B **98**, 075142 (2018)
63. Q. Wang, Y.-L. Zhu, Quenched Fano effect due to one Majorana zero mode coupled to the Fano interferometer. Curr. Appl. Phys. **18**(11), 1275–1279 (2018)
64. X.-Q. Wang, S.-F. Zhang, C. Jiang, G.-Y. Yi, W.-J. Gong, Fano antiresonance realized by Majorana bound states coupled to one quantum-dot system. Physica E **104**, 1–5 (2018)
65. A.M. Calle, M. Pacheco, P.A. Orellana, J.A. Otálora, Fano-Andreev and Fano-Majorana correspondence in quantum dot hybrid structures. Ann. Phys. **532**, 1900409 (2020)
66. A.P. Garrido, D. Zambrano, J.P. Ramos-Andrade, P.A. Orellana, Bound states in the continuum and Majorana zero modes in a double quantum dot interferometer: Ghost-Fano Majorana effect. Eur. Phys. J. Plus **138**, 663 (2023)
67. M. Aslam, D. Sivaganesh, S.Z. Naeimi, H. Shababi, M.K. Masood, A. Zatsepin, S. Arunachalam, Resonance and anti-resonance in transport through quantum dots-Majorana bound states hybrid structure. Physica E **158**, 115903 (2024)
68. H.-Z. Tang, Y.-T. Zhang, J.-J. Liu, Photon-assisted tunneling through a topological superconductor with Majorana bound states. AIP Adv. **5**(12), 127129 (2015)
69. F. Chi, T.-Y. He, J. Wang, Z.-G. Fu, L.-M. Liu, P. Liu, P. Zhang, Photon-assisted transport through a quantum dot side-coupled to Majorana bound states. Front. Phys. (Lausanne) **8**, 254 (2020)
70. T.-Y. He, H. Sun, G. Zhou, Photon-assisted seebeck effect in a quantum dot coupled to Majorana zero modes. Front. Phys. (Lausanne) **9**, 687438 (2021)
71. L.S. Ricco, V.K. Kozin, A.C. Seridonio, I.A. Shelykh, Reshaping the Jaynes-Cummings ladder with Majorana bound states. Phys. Rev. A **106**, 023702 (2022)

72. L.S. Ricco, V.K. Kozin, A.C. Seridonio, I.A. Shelykh, Accessing the degree of Majorana nonlocality in a quantum dot-optical microcavity system. Sci. Rep. **12**, 1983 (2022)
73. D. van Zanten, D. Sabonis, J. Suter, J.I. Väyrynen, T. Karzig, D.I. Pikulin, E. O'Farrell, D. Razmadze, K.D. Petersson, P. Krogstrup, C.M. Marcus, Photon-assisted tunnelling of zero modes in a Majorana wire. Nat. Phys. **16**, 663 (2020)
74. A.M. Calle, M. Pacheco, G.B. Martins, V.M. Apel, G.A. Lara, P.A. Orellana, Fano–Andreev effect in a T-shape double quantum dot in the Kondo regime. J. Phys.: Condens. Matter. **29**, 135301 (2017)
75. Y.-L. Huang, W.-K. Zou, H.-K. Zhao, Photon-assisted mesoscopic transport in a toroidal carbon nanotube coupled with Majorana bound states. Ann. Phys. **419**, 168239 (2020)
76. Y.-L. Huang, H.-K. Zhao, Differential conductance and shot noise of a graphene quantum dot coupled with Majorana bound states. Eur. Phys. J. B **92**, 256 (2019)
77. L. Fu, Electron teleportation via Majorana bound states in a mesoscopic superconductor. Phys. Rev. Lett. **104**, 056402 (2010)
78. Z. Wang, X.-Y. Hu, Q.-F. Liang, X. Hu, Detecting Majorana fermions by nonlocal entanglement between quantum dots. Phys. Rev. B **87**, 214513 (2013)
79. P. Wang, Y. Cao, M. Gong, G. Xiong, X.-Q. Li, Cross-correlations mediated by Majorana bound states. Europhys. Lett. **103**, 57016 (2013)
80. J. Nilsson, A.R. Akhmerov, C. Beenakker, Splitting of a Cooper pair by a pair of Majorana bound states. Phys. Rev. Lett. **101**, 120403 (2008)
81. B. Zocher, B. Rosenow, Modulation of Majorana-induced current cross-correlations by quantum dots. Phys. Rev. Lett. **111**, 036802 (2013)
82. J. Liu, F.-C. Zhang, K.T. Law, Majorana fermion induced nonlocal current correlations in spin-orbit coupled superconducting wires. Phys. Rev. B **88**, 064509 (2013)
83. J. Liu, J. Wang, F.-C. Zhang, Controllable nonlocal transport of Majorana fermions with the aid of two quantum dots. Phys. Rev. B **90**, 035307 (2014)
84. L. Hofstetter, S. Csonka, J. Nygård, C. Schnenberger, Cooper pair splitter realized in a two-quantum-dot Y-junction. Nature **461**, 960–963 (2009)
85. J.D. Sau, S.D. Sarma, Realizing a robust practical Majorana chain in a quantum-dot-superconductor linear array. Nat. Commun. **3**, 964 (2012)
86. M. Leijnse, K. Flensberg, Parity qubits and poor man's Majorana bound states in double quantum dots. Phys. Rev. B **86**, 134528 (2012)
87. I.C. Fulga, A. Haim, A.R. Akhmerov, Y. Oreg, Adaptive tuning of Majorana fermions in a quantum dot chain. New J. Phys. **15**, 045020 (2013)
88. A. Tsintzis, R.S. Souto, M. Leijnse, Creating and detecting poor man's Majorana bound states in interacting quantum dots. Phys. Rev. B **106**, L201404 (2022)
89. R.S. Souto, A. Tsintzis, M. Leijnse, J. Danon, Probing Majorana localization in minimal Kitaev chains through a quantum dot. Phys. Rev. Res. **5**, 043182 (2023)
90. A. Tsintzis, R.S. Souto, K. Flensberg, J. Danon, M. Leijnse, Majorana qubits and non-Abelian physics in quantum dot-based minimal Kitaev chains. PRX Quantum **5**, 010323 (2024)
91. T. Dvir, G. Wang, N. van Loo, C.-X. Liu, G.P. Mazur, A. Bordin, S. ten Haaf, J.-Y. Wang, D. van Driel, F. Zatelli, X. Li, F.K. Malinowski, S. Gazibegovic, G. Badawy, E. Bakkers, M. Wimmer, L.P. Kouwenhoven, Realization of a minimal Kitaev chain in coupled quantum dots. Nature **614**, 445–450 (2023)
92. S. ten Haaf, Q. Wang, A.M. Bozkurt, C.-X. Liu, I. Kulesh, P. Kim, D. Xiao, C. Thomas, M.J. Manfra, T. Dvir, M. Wimmer, S. Goswami, A two-site Kitaev chain in a two-dimensional electron gas. Nature **630**, 329–334 (2024)
93. F. Zatelli, D. van Driel, D. Xu, G. Wang, C.-X. Liu, A. Bordin, B. Roovers, G.P. Mazur, N. van Loo, J.C. Wolff, A.M. Bozkurt, G. Badawy, S. Gazibegovic, E. P. A. M. Bakkers, M. Wimmer, L.P. Kouwenhoven, T. Dvir, Robust poor man's Majorana zero modes using Yu-Shiba-Rusinov states. Nat. Commun. **15** (2024)
94. A. Bordin, C.-X. Liu, T. Dvir, F. Zatelli, S.L.D. ten Haaf, D. van Driel, G. Wang, N. van Loo, T. van Caekenberghe, J.C. Wolff, Y. Zhang, G. Badawy, S. Gazibegovic, E.P. A.M. Bakkers, M. Wimmer, L.P. Kouwenhoven, G.P. Mazur, Signatures of Majorana protection in a three-site Kitaev chain (2024)

95. Y.-X. Li, Z.-M. Bai, Tunneling transport through multi-quantum-dot with Majorana bound states. J. Appl. Phys. **114**(3), 033703 (2013)
96. W.-J. Gong, S.-F. Zhang, Z.-C. Li, G. Yi, Y.-S. Zheng, Detection of a Majorana fermion zero mode by a T-shaped quantum-dot structure. Phys. Rev. B **89**, 245413 (2014)
97. W.-J. Gong, S.-F. Zhang, Z.-C. Li, G. Yi, Y.-S. Zheng, Andreev reflection in a T-shaped double-quantum-dot structure induced by Majorana bound states. J. Phys. Soc. Jpn. **83**(3), 034706 (2014)
98. C. Jiang, G. Lu, W.-J. Gong, Tunable transport through a quantum dot chain with side-coupled Majorana bound states. J. Appl. Phys. **116**(10), 103704 (2014)
99. J.F. Silva, E. Vernek, Andreev and Majorana bound states in single and double quantum dot structures. J. Phys.: Condens. Matter. **28**, 435702 (2016)
100. Z.-T. Jiang, Z.-Y. Cao, C.-C. Zhong, Quantum transport through a triple quantum dot system in the presence of Majorana bound states. Commun. Theor. Phys. **65**, 622–628 (2016)
101. T.I. Ivanov, Coherent tunneling through a double quantum dot coupled to Majorana bound states. Phys. Rev. B **96**, 035417 (2017)
102. J.D. Cifuentes, L. da Silva, Manipulating Majorana zero modes in double quantum dots. Phys. Rev. B **100**, 085429 (2019)
103. V. Mourik, K. Zuo, S.M. Frolov, S.R. Plissard, E. Bakkers, L.P. Kouwenhoven, Signatures of Majorana fermions in hybrid superconductor-semiconductor nanowire devices. Science **336**(6084), 1003 (2012)
104. A. Das, Y. Ronen, Y. Most, Y. Oreg, M. Heiblum, H. Shtrikman, Zero-bias peaks and splitting in an Al–InAs nanowire topological superconductor as a signature of Majorana fermions. Nature Phys. **8**, 887–895 (2012)
105. M.T. Deng, C.L. Yu, G.Y. Huang, M. Larsson, P. Caroff, H.Q. Xu, Anomalous zero-bias conductance peak in a Nb–InSb nanowire–Nb hybrid device. Nano Lett. **12**, 6414–6419 (2012)
106. A. Finck, D.J. Van Harlingen, P.K. Mohseni, K. Jung, X. Li, Anomalous modulation of a zero-bias peak in a hybrid nanowire-superconductor device. Phys. Rev. Lett. **110**, 126406 (2013)
107. H. Churchill, V. Fatemi, K. Grove-Rasmussen, M.T. Deng, P. Caroff, H.Q. Xu, C.M. Marcus, Superconductor-nanowire devices from tunneling to the multichannel regime: Zero-bias oscillations and magnetoconductance crossover. Phys. Rev. B **87**, 241401 (2013)
108. S.M. Albrecht, A.P. Higginbotham, M. Madsen, F. Kuemmeth, T.S. Jespersen, J. Nygrd, P. Krogstrup, C.M. Marcus, Exponential protection of zero modes in Majorana islands. Nature **531**, 206–209 (2016)
109. J. Chen, P. Yu, J. Stenger, M. Hocevar, D. Car, S.R. Plissard, E. Bakkers, T.D. Stanescu, S.M. Frolov, Experimental phase diagram of zero-bias conductance peaks in superconductor/semiconductor nanowire devices. Sci. Adv. **3**, e1701476 (2017)
110. H. Zhang, O. Gül, S. Conesa-Boj, M.P. Nowak, M. Wimmer, K. Zuo, V. Mourik, F.K. de Vries, J. van Veen, M. de Moor, J. Bommer, D.J. van Woerkom, D. Car, S.R. Plissard, E. Bakkers, M. Quintero-Prez, M.C. Cassidy, S. Koelling, S. Goswami, K. Watanabe, T. Taniguchi, L.P. Kouwenhoven, Ballistic superconductivity in semiconductor nanowires. Nat. Commun. **8**(1), 16025 (2017)
111. J. Kammhuber, M.C. Cassidy, F. Pei, M.P. Nowak, A. Vuik, Ö. Gül, D. Car, S.R. Plissard, E.P. A.M. Bakkers, M. Wimmer, L.P. Kouwenhoven, Conductance through a helical state in an Indium antimonide nanowire. Nat. Commun. **8** (2017)
112. F. Nichele, A. Drachmann, A.M. Whiticar, E. O'Farrell, H.J. Suominen, A. Fornieri, T. Wang, G.C. Gardner, C. Thomas, A.T. Hatke, P. Krogstrup, M.J. Manfra, K. Flensberg, C.M. Marcus, Scaling of Majorana zero-bias conductance peaks. Phys. Rev. Lett. **119**, 136803 (2017)
113. O. Gl, H. Zhang, J. Bommer, M. de Moor, D. Car, S.R. Plissard, E. Bakkers, A. Geresdi, K. Watanabe, T. Taniguchi, L.P. Kouwenhoven, Ballistic Majorana nanowire devices. Nat. Nanotechnol. **13**, 192–197 (2018)
114. R.M. Lutchyn, E. Bakkers, L.P. Kouwenhoven, P. Krogstrup, C.M. Marcus, Y. Oreg, Majorana zero modes in superconductor–semiconductor heterostructures. Nat. Rev. Mater. **3**, 52–68 (2018)

115. S. Vaitiekėnas, M.-T. Deng, J. Nygård, P. Krogstrup, C.M. Marcus, Effective *g* factor of subgap states in hybrid nanowires. Phys. Rev. Lett. **121**, 037703 (2018)
116. A. Grivnin, E. Bor, M. Heiblum, Y. Oreg, H. Shtrikman, Concomitant opening of a bulk-gap with an emerging possible Majorana zero mode. Nat. Commun. **10** (2019)
117. J.-Y. Wang, N. van Loo, G.P. Mazur, V. Levajac, F.K. Malinowski, M. Lemang, F. Borsoi, G. Badawy, S. Gazibegovic, E. Bakkers, M. Quintero-Pérez, S. Heedt, L.P. Kouwenhoven, Parametric exploration of zero-energy modes in three-terminal InSb-Al nanowire devices. Phys. Rev. B **106**, 075306 (2022)
118. Z. Wang, H. Song, D. Pan, Z. Zhang, W. Miao, R. Li, Z. Cao, G. Zhang, L. Liu, L. Wen, R. Zhuo, D.E. Liu, K. He, R. Shang, J. Zhao, H. Zhang, Plateau regions for zero-bias peaks within 5 $2e^2/h$. Phys. Rev. Lett. **129**, 167702 (2022)
119. M. Aghaee, A. Akkala, Z. Alam, R. Ali, A. Alcaraz Ramirez, M. Andrzejczuk, A. E. Antipov, P. Aseev, M. Astafev, B. Bauer, J. Becker, S. Boddapati, F. Boekhout, J. Bommer, T. Bosma, L. Bourdet, S. Boutin, P. Caroff, L. Casparis, M. Cassidy, S. Chatoor, A. W. Christensen, N. Clay, W. S. Cole, F. Corsetti, A. Cui, P. Dalampiras, A. Dokania, G. de Lange, M. de Moor, J. C. Estrada Saldaña, S. Fallahi, Z. H. Fathabad, J. Gamble, G. Gardner, D. Govender, F. Griggio, R. Grigoryan, S. Gronin, J. Gukelberger, E. B. Hansen, S. Heedt, J. Herranz Zamorano, S. Ho, U. L. Holgaard, H. Ingerslev, L. Johansson, J. Jones, R. Kallaher, F. Karimi, T. Karzig, E. King, M. E. Kloster, C. Knapp, D. Kocon, J. Koski, P. Kostamo, P. Krogstrup, M. Kumar, T. Laeven, T. Larsen, K. Li, T. Lindemann, J. Love, R. Lutchyn, M. H. Madsen, M. Manfra, S. Markussen, E. Martinez, R. McNeil, E. Memisevic, T. Morgan, A. Mullally, C. Nayak, J. Nielsen, W. H. P. Nielsen, B. Nijholt, A. Nurmohamed, E. O'Farrell, K. Otani, S. Pauka, K. Petersson, L. Petit, D. I. Pikulin, F. Preiss, M. Quintero-Perez, M. Rajpalke, K. Rasmussen, D. Razmadze, O. Reentila, D. Reilly, R. Rouse, I. Sadovskyy, L. Sainiemi, S. Schreppler, V. Sidorkin, A. Singh, S. Singh, S. Sinha, P. Sohr, T. c. v. Stankevič, L. Stek, H. Suominen, J. Suter, V. Svidenko, S. Teicher, M. Temuerhan, N. Thiyagarajah, R. Tholapi, M. Thomas, E. Toomey, S. Upadhyay, I. Urban, S. Vaitiekėnas, K. Van Hoogdalem, D. Van Woerkom, D. V. Viazmitinov, D. Vogel, S. Waddy, J. Watson, J. Weston, G. W. Winkler, C. K. Yang, S. Yau, D. Yi, E. Yucelen, A. Webster, R. Zeisel, and R. Zhao, "InAs-Al hybrid devices passing the topological gap protocol," *Phys. Rev. B*, vol. 107, p. 245423, 2023
120. Y. Wang, F. Chen, W. Song, Z. Geng, Z. Yu, L. Yang, Y. Gao, R. Li, S. Yang, W. Miao, W. Xu, Z. Wang, Z. Xia, H.-D. Song, X. Feng, T. Wang, Y. Zang, L. Li, R. Shang, Q. Xue, K. He, H. Zhang, Ballistic PbTe nanowire devices. Nano Lett. **23**, 11137–11144 (2023)
121. Y. Gao, W. Song, Z. Yu, S. Yang, Y. Wang, R. Li, F. Chen, Z. Geng, L. Yang, J. Xu, Z. Wang, Z. Li, S. Zhang, X. Feng, T. Wang, Y. Zang, L. Li, R. Shang, Q.-K. Xue, K. He, H. Zhang, SQUID oscillations in PbTe nanowire networks. Phys. Rev. B **110**, 045405 (2024)
122. R. Li, W. Song, W. Miao, Z. Yu, Z. Wang, S. Yang, Y. Gao, Y. Wang, F. Chen, Z. Geng, L. Yang, J. Xu, X. Feng, T. Wang, Y. Zang, L. Li, R. Shang, Q. Xue, K. He, H. Zhang, Selective-area-grown PbTe-Pb planar Josephson junctions for quantum devices. Nano Lett. **24**, 4658–4664 (2024)
123. M.T. Deng, S. Vaitiekenas, E.B. Hansen, J. Danon, M. Leijnse, K. Flensberg, J. Nygård, P. Krogstrup, C.M. Marcus, Majorana bound state in a coupled quantum-dot hybrid-nanowire system. Science **354**(6319), 1557–1562 (2016)
124. D. Sherman, J.S. Yodh, S.M. Albrecht, J. Nygård, P. Krogstrup, C.M. Marcus, Normal, superconducting and topological regimes of hybrid double quantum dots. Nat. Nanotechnol. **12**, 212–217 (2017)
125. M.-T. Deng, S. Vaitiekėnas, E. Prada, P. San-Jose, J. Nygård, P. Krogstrup, R. Aguado, C.M. Marcus, Nonlocality of Majorana modes in hybrid nanowires. Phys. Rev. B **98**, 085125 (2018)
126. P.P. Aseev, J. Klinovaja, D. Loss, Lifetime of Majorana qubits in Rashba nanowires with nonuniform chemical potential. Phys. Rev. B **98**, 155414 (2018)
127. P.P. Aseev, P. Marra, P. Stano, J. Klinovaja, D. Loss, Degeneracy lifting of Majorana bound states due to electron-phonon interactions. Phys. Rev. B **99**, 205435 (2019)
128. Z.-T. Zhang, F. Mei, X.-G. Meng, B.-L. Liang, Z.-S. Yang, Effects of decoherence on diabatic errors in Majorana braiding. Phys. Rev. A **100**, 012324 (2019)

129. J. Schulenborg, M. Burrello, M. Leijnse, K. Flensberg, Multilevel effects in quantum dot based parity-to-charge conversion of Majorana box qubits. Phys. Rev. B **103**, 245407 (2021)
130. G. Goldstein, C. Chamon, Decay rates for topological memories encoded with Majorana fermions. Phys. Rev. B **84**, 205109 (2011)
131. T.L. Schmidt, A. Nunnenkamp, C. Bruder, Majorana qubit rotations in microwave cavities. Phys. Rev. Lett. **110**, 107006 (2013)
132. T.L. Schmidt, A. Nunnenkamp, C. Bruder, Microwave-controlled coupling of Majorana bound states. New J. Phys. **15**, 025043 (2013)
133. F.L. Pedrocchi, D.P. DiVincenzo, Majorana braiding with thermal noise. Phys. Rev. Lett. **115**, 120402 (2015)
134. C. Knapp, T. Karzig, R.M. Lutchyn, C. Nayak, Dephasing of Majorana-based qubits. Phys. Rev. B **97**, 125404 (2018)
135. Y. Song, S. Das Sarma, Phonon-induced Majorana qubit relaxation in tunnel-coupled two-island topological superconductors. Phys. Rev. B **98**, 075159 (2018)
136. H.-L. Lai, P.-Y. Yang, Y.-W. Huang, W.-M. Zhang, Exact master equation and non-Markovian decoherence dynamics of Majorana zero modes under gate-induced charge fluctuations. Phys. Rev. B **97**, 054508 (2018)
137. J.C. Budich, S. Walter, B. Trauzettel, Failure of protection of Majorana based qubits against decoherence. Phys. Rev. B **85**, 121405 (2012)
138. D. Rainis, D. Loss, Majorana qubit decoherence by quasiparticle poisoning. Phys. Rev. B **85**, 174533 (2012)
139. A.E. Antipov, A. Bargerbos, G.W. Winkler, B. Bauer, E. Rossi, R.M. Lutchyn, Effects of gate-induced electric fields on semiconductor Majorana nanowires. Phys. Rev. X **8**, 031041 (2018)
140. M.J. Schmidt, D. Rainis, D. Loss, Decoherence of Majorana qubits by noisy gates. Phys. Rev. B **86**, 085414 (2012)
141. F. Setiawan, J.D. Sau, Electron-boson-interaction induced particle-hole symmetry breaking of conductance into subgap states in superconductors. Phys. Rev. Res. **3**, L032038 (2021)
142. S.-N. Zhang, W. Pei, T.-F. Fang, Q.-F. Sun, Phonon-assisted transport through quantum dots with normal and superconducting leads. Phys. Rev. B **86**, 104513 (2012)
143. J. Barański, T. Domański, Fano-type resonances induced by a boson mode in Andreev conductance. Chin. Phys. B **24**, 017304 (2015)
144. J. Barański, T. Domański, Enhancements of the Andreev conductance due to emission/absorption of bosonic quanta. J. Phys.: Condens. Matter. **27**, 305302 (2015)
145. P. Stadler, W. Belzig, G. Rastelli, Ground-state cooling of a mechanical oscillator by interference in Andreev reflection. Phys. Rev. Lett. **117**, 197202 (2016)
146. P. Stadler, W. Belzig, G. Rastelli, Charge-vibration interaction effects in normal-superconductor quantum dots. Phys. Rev. B **96**, 045429 (2017)
147. Z. Cao, T.-F. Fang, Q.-F. Sun, H.-G. Luo, Inelastic Kondo-Andreev tunneling in a vibrating quantum dot. Phys. Rev. B **95**, 121110 (2017)
148. B. Dong, G.H. Ding, X.L. Lei, Full counting statistics of phonon-assisted Andreev tunneling through a quantum dot coupled to normal and superconducting leads. Phys. Rev. B **95**, 035409 (2017)
149. P. Zhang, Y. xian Li, Andreev reflection and shot noise in a quantum dot with phonon modes. J. Phys.: Condens. Matter. **21**, 095602 (2009)
150. L. Bai, Z.-Z. Zhang, L. Jiang, Andreev reflection current through a molecule quantum dot in the presence of the electron-phonon interaction and the spin-flip scattering. Phys. Lett. A **375**, 661–665 (2011)
151. K. Bocian, W. Rudziński, Andreev reflection in transport through a quantum dot interacting with a polaron. Acta Phys. Pol., A **126**, 374–375 (2014)
152. K. Bocian, W. Rudziński, Phonon-assisted Andreev reflection in a hybrid junction based on a quantum dot. Eur. Phys. J. B **88**, 50 (2015)
153. J. Gramich, A. Baumgartner, C. Schönenberger, Resonant and inelastic Andreev tunneling observed on a carbon nanotube quantum dot. Phys. Rev. Lett. **115**, 216801 (2015)

154. N. Dai, Q.-F. Sun, Phonon-assisted Andreev reflection at a Majorana zero mode. Phys. Rev. B **99**, 085436 (2019)
155. X.-Q. Wang, B. Wu, S.-F. Zhang, Q. Wang, W.-J. Gong, Influences of electron-phonon interaction on quantum transport through one quantum-dot system with side-coupled Majorana zero mode. Ann. Phys. **415**, 168127 (2020)
156. X.-D. Wang, X.-Q. Wang, H.-Y. Sun, X.-F. Dai, Q. Wang, W.-J. Gong, Phonon-assisted Andreev reflection driven by a Majorana zero mode. Phys. Scr. **96**, 015805 (2021)
157. Z.-H. Wang, Heat generation by electrical current in a quantum dot hybridized to Majorana nanowire. Front. Phys. (Lausanne) **9**, 704493 (2021)
158. Z.-H. Wang, W.-C. Huang, Dual negative differential of heat generation in a strongly correlated quantum dot side-coupled to Majorana bound states. Front. Phys. (Lausanne) **9**, 727934 (2021)
159. L. Máthé, D. Sticlet, L.P. Zârbo, Quantum transport through a quantum dot side-coupled to a Majorana bound state pair in the presence of electron-phonon interaction. Phys. Rev. B **105**, 155409 (2022)
160. L. Máthé, Z. Kovács-Krausz, I. Botiz, I. Grosu, K. El Anouz, A. El Allati, L.P. Zârbo, Phonon-assisted tunneling through quantum dot systems connected to majorana bound states. Nanomaterials **13**(10) (2023)
161. J.-X. Zhu, A.V. Balatsky, Theory of current and shot-noise spectroscopy in single-molecular quantum dots with a phonon mode. Phys. Rev. B **67**, 165326 (2003)
162. Z.-Z. Chen, R. Lü, B.-F. Zhu, Effects of electron-phonon interaction on nonequilibrium transport through a single-molecule transistor. Phys. Rev. B **71**, 165324 (2005)
163. Y. Meir, N.S. Wingreen, P.A. Lee, Transport through a strongly interacting electron system: Theory of periodic conductance oscillations. Phys. Rev. Lett. **66**, 3048–3051 (1991)
164. Y. Meir, N.S. Wingreen, P.A. Lee, Low-temperature transport through a quantum dot: The Anderson model out of equilibrium. Phys. Rev. Lett. **70**, 2601–2604 (1993)
165. N.S. Wingreen, Y. Meir, Anderson model out of equilibrium: Noncrossing-approximation approach to transport through a quantum dot. Phys. Rev. B **49**, 11040–11052 (1994)
166. A.-P. Jauho, N.S. Wingreen, Y. Meir, Time-dependent transport in interacting and noninteracting resonant-tunneling systems. Phys. Rev. B **50**, 5528–5544 (1994)
167. Q.-F. Sun, B.-G. Wang, J. Wang, T.-H. Lin, Electron transport through a mesoscopic hybrid multiterminal resonant-tunneling system. Phys. Rev. B **61**, 4754–4761 (2000)
168. G. Michałek, B.R. Bułka, T. Domański, K.I. Wysokiński, Interplay between direct and crossed Andreev reflections in hybrid nanostructures. Phys. Rev. B **88**, 155425 (2013)
169. G. Michałek, T. Domański, B. Bułka, K. Wysokiński, Novel non-local effects in three-terminal hybrid devices with quantum dot. Sci. Rep. **5**, 14572 (2015)
170. K. Bocian, W. Rudziński, Photon-assisted tunneling in a hybrid junction based on a quantum dot coupled to two ferromagnets and a superconductor. Phys. Status Solidi B **254**, 1600206 (2016)
171. S.-Y. Hwang, R. López, D. Sánchez, Cross thermoelectric coupling in normal-superconductor quantum dots. Phys. Rev. B **91**, 104518 (2015)
172. M.B. Tagani, H.R. Soleimani, Phonon-assisted tunneling through a double quantum dot system. Phys. Scr. **86**, 035706 (2012)
173. Z.-Z. Chen, H. Lu, R. L, B. fen Zhu, Phonon-assisted Kondo effect in a single-molecule transistor out of equilibrium. J. Phys.: Condens. Matter. **18,** 5435 (2006)
174. F. Jiang, H. Xie, Y. Yan, Spin-dependent thermoelectric effect in polaronic Kondo transport through a Rashba quantum dot coupled with ferromagnetic leads. Phys. Lett. A **378**(26), 1854–1866 (2014)
175. K.T. Law, P.A. Lee, T.K. Ng, Majorana fermion induced resonant Andreev reflection. Phys. Rev. Lett. **103**, 237001 (2009)
176. G.D. Mahan, *Many-Particle Physics*, 3rd edn. (Plenum, New York, 2000)
177. R. Świrkowicz, M. Wilczyński, J. Barnaś, The Kondo effect in quantum dots coupled to ferromagnetic leads with noncollinear magnetizations: Effects due to electron-phonon coupling. J. Phys.: Condens. Matter. **20**, 255219 (2008)

178. K.-H. Yang, C.-Z. Liu, AC shot noise through a vibrating quantum dot. Eur. Phy. J. B **72**, 409–415 (2009)
179. L. Máthé, I. Grosu, Transport through a quantum dot with electron-phonon interaction. Mater. Today: Proc. **5**(8), 15878–15887 (2018)
180. C. Lacroix, Density of states for the Anderson model. J. Phys. F: Met. Phys. **11**, 2389–2397 (1981)
181. V. Kashcheyevs, A. Aharony, O. Entin-Wohlman, Applicability of the equations-of-motion technique for quantum dots. Phys. Rev. B **73**, 125338 (2006)
182. R. Van Roermund, S.-Y. Shiau, M. Lavagna, Anderson model out of equilibrium: Decoherence effects in transport through a quantum dot. Phys. Rev. B **81**, 165115 (2010)
183. U. Lundin, R.H. McKenzie, Temperature dependence of polaronic transport through single molecules and quantum dots. Phys. Rev. B **66**, 075303 (2002)
184. S. Sapmaz, P. Jarillo-Herrero, Y.M. Blanter, C. Dekker, H. van der Zant, Tunneling in suspended carbon nanotubes assisted by longitudinal phonons. Phys. Rev. Lett. **96**, 026801 (2006)
185. R. Leturcq, C. Stampfer, K. Inderbitzin, L. Durrer, C. Hierold, E. Mariani, M.G. Schultz, F. von Oppen, K. Ensslin, Franck–Condon blockade in suspended carbon nanotube quantum dots. Nat. Phys. **5**, 327–331 (2009)
186. G. Górski, J.A. Barański, I. Weymann, T. Domański, Interplay between correlations and Majorana mode in proximitized quantum dot. Sci. Rep. **8**, 15717 (2018)
187. Q.-F. Sun, X.C. Xie, Heat generation by electric current in mesoscopic devices. Phys. Rev. B **75**, 155306 (2007)
188. N.A. Zimbovskaya, Electron transport through a quantum dot in the Coulomb blockade regime: Nonequilibrium Green's function based model. Phys. Rev. B **78**, 035331 (2008)
189. L.-L. Zhou, X.-Y. Zhou, R. Cheng, Electron tunneling through a coupled system of electron and phonon. Chin. J. Phys. **60**, 307–312 (2019)
190. D. Razmadze, E. O'Farrell, P. Krogstrup, C.M. Marcus, Quantum dot parity effects in trivial and topological Josephson junctions. Phys. Rev. Lett. **125**, 116803 (2020)
191. L. Máthé, I. Grosu, Nonequilibrium Kondo effect in a graphene-coupled quantum dot in the presence of a magnetic field. Beilstein J. Nanotechnol. **11**, 225–239 (2020)

Chapter 3
Nonequilibrium Kondo Physics of a Quantum Dot Connected to Graphene Leads in the Presence of Magnetic Fields

This chapter of the book focuses on the theoretical study of quantum transport in a setup involving the embedding of a semiconductor quantum dot (QD) between two monolayer graphene leads. The interacting QD is threaded by an external magnetic field, resulting in spin-polarized quantum transport. The proposed system provides an ideal platform for investigating many-body quantum phenomena, such as Kondo-type transport, which is treated here within the framework of the nonequilibrium Green's function formalism. Additionally, the device under investigation can serve as a reference for the development of graphene-based nanoelectronic devices, leveraging the unique electronic properties of the graphene.

3.1 State-of-The-Art and Motivation

The discovery of graphene [1, 2] has extended the existing research areas and allowed the development of novel quantum nanodevices based on graphene [3–6] which reveal unique physical properties [7–9]. As mentioned in Sect. 1.1.1 such quantum devices are nanojunctions whose implementation can based on the coupling of QDs to much larger systems like metallic [10–30], ferromagnetic [31–34], or graphene leads [35–39]. The choice of electrodes can significantly influence the transport properties of the QD-based setup.

Low-dimensional quantum systems, such as QDs connected to charge reservoirs, can be described by the Anderson impurity model [40], for which the Kondo effect can be seen within an appropriate approximation, e.g., within the Lacroix approximation [41]. The Kondo effect, discussed in Sect. 1.1.2, is a result of the interaction of a magnetic impurity (e.g., a localized spin) with spin-dependent conduction electrons [42, 43]. In the past few decades, the Kondo problem in QDs has attracted much attention in scientific community. In such a way, the Kondo effect in QDs connected to normal or ferromagnetic leads has been extensively analyzed theoretically by using

© The Author(s), under exclusive license to Springer Nature Switzerland AG 2025

L. Máthé, *Physical Properties of Low-Dimensional Systems*,
NanoScience and Technology, https://doi.org/10.1007/978-3-031-91602-1_3

the Anderson impurity model within the equation of motion (EOM) framework [13, 19, 20, 23, 25, 28, 29, 31, 44–46]. The noncrossing approximation (NCA) method was used to investigate the nonequilibrium Kondo problem in QDs in the absence of magnetic fields within the single-impurity Anderson model [15, 19, 47–51]. In addition, the transport characteristics of QDs have also been analyzed within the framework of modified perturbation theory [12, 14, 21], numerical renormalization group (NRG) approach [51–55], quantum Monte Carlo (QMC) method [56, 57] and slave-boson mean-field (SBMF) theory [17, 58, 59], respectively. It was found that Kondo resonances show up in the nonequilibrium density of states (DOS) of QD at the chemical potentials of leads and the height of peaks is suppressed under dissipation processes [13]. The approximation scheme proposed by Meir and co-workers [10] was applied to study the Kondo physics appearing in strongly interacting QDs [13, 45, 46] or in QD systems with finite Coulomb interaction [20, 23, 25]. The Kondo physics of nonequilibrium QDs in the presence of a magnetic field has been studied within the EOM [13], NCA [13], NRG [60, 61], and SBMF [18] methods. It was established that the applied magnetic field on QD leads to the splitting of Kondo resonances in the DOS of QD. Namely, the peaks which emerge originally at the chemical potentials without magnetic fields move by the Zeeman splitting to the left in energy (lower energies) for electrons with spin-down and to the right (higher energies) for electrons with spin-up. Furthermore, the Kondo peak develops in the differential conductance plot when the applied voltage to the leads equals the Zeeman splitting. Numerous experimental efforts have been made to confirm the Kondo effect in such systems [24, 42, 43, 60, 62–76].

The effects of magnetic adatoms and the presence of vacancies on the physical properties of graphene have been theoretically investigated in the literature [77–83]. It was established that magnetic moments can be created by electrically tuning the properties of transition metal adatoms on graphene surface providing thus the possibility for application of graphene in the development of spintronic devices [79]. Furthermore, several works have studied the physical properties of such systems using density functional theory (DFT), where the graphene is doped with impurities or graphene with vacancies is considered [84–92]. In addition, the effects of adatom adsorption on graphene's electronic structure and transport properties were investigated using the SBMF approach [93] and the T-matrix approximation [94–96], respectively. Various theoretical works treat the Kondo effect in graphene within magnetic impurity framework described by a pseudogap Anderson model [93, 97–102]. The Kondo effect was analyzed for different positions of a magnetic impurity on graphene or graphene with vacancies within the tight-binding formalism [99]. Zhu and Berakdar explored theoretically the DOS in the Kondo regime of a system composed by the deposition of a magnetic adatom on graphene employing the EOM method within Lacroix's approximation [100]. They established that the Kondo peak emerges only in a narrow energy window of the impurity level measured from the chemical potential. Ref. [103] reports theoretical calculations regarding the Kondo effect within the nonequilibrium Green's function (NGF) formalism in an adatom deposited on the surface of a graphene and its scanning tunneling microscopy (STM). It has been established that the tunneling channel between the graphene and STM tip has insignificant influence on the lineshape of Kondo peak around the zero-bias regime. In addition, Yanagisawa [104] analyzed the Kondo problem in a Dirac system for which the localized spin is coupled to Dirac electrons within the framework of Green's function formalism. It was obtained that the Kondo temperature reveals a proportionally dependence on the chemical potential $|\mu|$ and the Kondo screening disappears when the chemical potential is at the Dirac point. The NRG approach was employed to investigate the Kondo effect for a magnetic impurity within gapless Kondo [105, 106] or Anderson models [106, 107]. It was established that the Kondo peak is not detected when the Fermi level is located at the Dirac point. Furthermore, the Kondo effect produced by a magnetic impurity or a point defect in graphene has been profoundly analyzed by using NRG approach [108–115], DFT calculations [115–118], SBMF approximation [119] and dynamical mean-field theory combined with QMC simulations [120]. May and coauthors investigated the Kondo effect for carbon point defects in monolayer graphene by using an effective two-orbital quantum impurity model via NRG technique [121]. In the study [122], the pseudogap Anderson and Kondo models were applied within a combination of numerical and analytical renormalization group methods to investigate the Kondo screening for magnetic impurities in doped graphene. Interestingly, they found that the Kondo temperature behaves differently depending on the chemical potential sign in the strong Coulomb interaction limit ($U \to \infty$). Details on the experimental investigation of Kondo effect in graphene induced by lattice vacancies or adatoms via transport characteristic measurements have been reported elsewhere [123–125]. Other experimental realizations of graphene doped with impurities or graphene with vacancies have been reported in the literature [126–136].

The vast majority of experimental measurements in recent years have attracted a lot of attention to the exploration of transport properties of graphene QDs [137–140], graphene QDs in the multi-level regime [141–143], bilayer graphene QDs [144, 145], graphene double QDs [146, 147], and graphene triple QDs [148] in the Coulomb-blockade regime. Up to now, less attention has been paid in the literature to analyze the transport phenomena in QDs connected to graphene systems within the pseudogap Anderson model. Therefore, a couple of theoretical works concerning the investigation of thermoelectric characteristics via NCA calculations in strongly interacting QDs coupled to leads of massless Dirac fermions have been reported in Refs. [35, 36]. It was found that there is an impurity induced quantum phase transition in the zero-bias conductance between the local moment and Kondo regimes, and the thermopower alters its sign at temperature near the Kondo temperature. Aono and Komine analyzed the thermoelectric characteristics of a noninteracting QD attached to massless Dirac fermions and found that, by tuning the QD energy, such systems can reach large values for the figure of merit and thermopower [37]. The thermoelectric transport in Coulomb-blockade regime of a dot connected to monolayer graphene leads has been studied employing the EOM method within the Hartree-Fock approximation [39]. They concluded that the Wiedemann-Franz law is not satisfied for graphene electrodes. The thermoelectric characteristics of a single QD attached to pure as well as to gapped graphene leads have been investigated applying the EOM technique within the Hartree-Fock approximation [38]. A remarkable amplification in the figure of merit was found for a QD system with gapped graphene leads within the massless gap scenario. Such kind of setups reveal high efficiency for heat-to-electricity conversion at low temperatures for which the contribution of phonons can be neglected [37, 39].

As we have seen above, the Kondo problem for magnetic impurities in graphene lattice was mainly studied in the infinite Coulomb interaction limit using analytical methods. The aim of this chapter is to investigate the nonequilibrium Kondo-type transport in a QD attached to graphene leads for which the dot with finite Coulomb energy is threaded by a magnetic field. The system is modeled by a pseudogap Anderson Hamiltonian [35–39, 93, 97, 100–102] within the NGF formalism [16, 149, 150]. The dot retarded Green's function is determined with the EOM technique [25, 39, 41, 45, 46, 100, 151]. We then derive a formula for the differential conductance and an approximated relation for the Kondo temperature for finite Coulomb interaction when the leads are differently doped. The findings for DOS and differential conductance in Kondo regime without and with magnetic field are discussed. The analytical determination of integrals appearing in self-energies, based on a simple method we developed, is presented in our article [151] and could be useful for further studies of graphene-based systems. We finally summarize the main findings of this chapter. Here, we briefly present our central results. We find that the Kondo temperature vanishes at the Dirac point and near the particle-hole symmetric point in dot. The influence of magnetic field and finite Coulomb interaction on the transport characteristics of the device shows similar tendency to results reported in the literature for QDs coupled to normal leads. It has been found that the Kondo peak does not develop in the DOS of QD when the chemical potentials of leads are pinned at the Dirac points in graphene due to the linear energy dispersion of massless Dirac fermions, in agreement with previous studies. Our QD-graphene setup leads to theoretically investigate the quantum many-body physics of Kondo effect and provides a possible platform for developing novel graphene-based nanodevices.

3.2 Theoretical Model and Analytical Results

3.2.1 Model Hamiltonian of the System

We consider a QD laterally connected to two leads realized from monolayer graphene as it is illustrated in Fig. 3.1. The system is modeled with a pseudogap Anderson Hamiltonian as [35–39, 93, 97, 100–102, 151]:

$$\mathcal{H} = \mathcal{H}_{\text{Leads}} + \mathcal{H}_{\text{QD}} + \mathcal{H}_{\text{T}}. \tag{3.1}$$

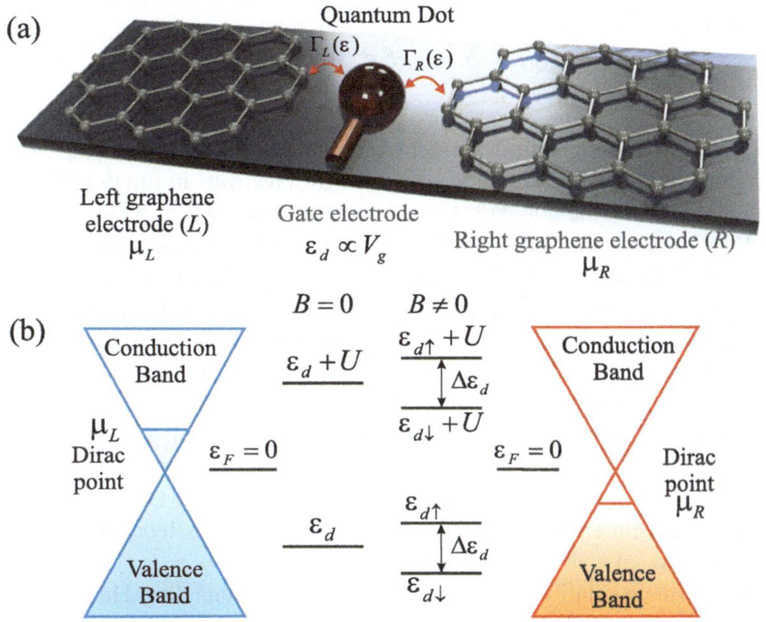

Fig. 3.1 **a** Schematic representation of a quantum dot (QD) connected to left (L) and right (R) monolayer graphene leads with chemical potentials (μ_L and μ_R). Here, $\Gamma_\alpha(\varepsilon)$ stands for the coupling strength between the dot and α graphene lead. The spin-independent dot energy level ε_d is tuned by the gate voltage V_g via the gate electrode. **b** The schematic energy level diagram of the QD with finite Coulomb interaction (U) in the absence ($B = 0$) and presence ($B \neq 0$) of a magnetic field when the chemical potentials of leads are tuned away from the Dirac point ($\mu_\alpha \neq \varepsilon_F$ with $\varepsilon_F = 0$). Reproduced from [151] under CC-BY 4.0 license. Copyright 2020 by the authors

The first term in (3.1), $\mathcal{H}_{\text{Leads}}$, symbolizes the Hamiltonian of graphene electrodes, which models the massless Dirac fermions in the leads, and it reads

$$\mathcal{H}_{\text{Leads}} = \sum_{\alpha,s,\sigma} \int_{-k_c}^{+k_c} dk\, \varepsilon_k\, c^\dagger_{\alpha s k \sigma} c_{\alpha s k \sigma}, \tag{3.2}$$

where $c_{\alpha s k \sigma}$ and $c^\dagger_{\alpha s k \sigma}$ represent the annihilation and creation operators for massless Dirac fermions with momentum k and spin σ in the graphene lead α [$\alpha \equiv$ left (L) and right (R)]. Here, s stands for the valley index and $\varepsilon_k = \hbar v_F k$ denotes the linear energy dispersion relation for massless Dirac fermions in graphene where the graphene Fermi velocity is $v_F \approx 10^6$ m/s. The graphene leads are at the same temperatures, i.e., $T_\alpha = T$. Therefore, $f_\alpha(\varepsilon_k) = 1/\left[e^{(\varepsilon_k - \mu_\alpha)/k_B T} + 1\right]$ is the Fermi-Dirac distribution function for massless Dirac fermions in α graphene lead with μ_α being the chemical potential that is considered temperature-independent. We define a momentum cutoff k_c such that the linear dispersion relation to be fulfilled, where $D = \hbar v_F k_c$ being the energy cutoff.

The second term in relation (3.1), $\mathcal{H}_{\text{QD}}$, describes the interacting QD and it is expressed as

$$\mathcal{H}_{\text{QD}} = \sum_{\sigma} \varepsilon_{d\sigma} d_{\sigma}^{\dagger} d_{\sigma} + U n_{d\uparrow} n_{d\downarrow}, \tag{3.3}$$

where $\varepsilon_{d\sigma}$ represents the spin-dependent QD energy level. Here, d_{σ} ($d_{\sigma}^{\dagger}$) stands for the fermionic annihilation (creation) operator for electrons in the dot. The applied magnetic field (B) on the dot leads to an energy splitting of the QD level, i.e., $\varepsilon_{d\sigma} = \varepsilon_d + \sigma \Delta\varepsilon_d/2$ where $\sigma = +1$ and $\bar{\sigma} = -1$ correspond to the electrons with spin up ($\uparrow$) and down ($\downarrow$), respectively. The QD level ε_d is tuned by the gate voltage V_g applied to the metallic gate electrode. Recall that $\Delta\varepsilon_d = |g|\mu_B B$ is the Zeeman splitting where g and μ_B are the Landé factor and the Bohr magneton, respectively. Therefore, the second term in Eq. (3.3) refers to the repulsion interaction of QD electrons with Coulomb energy U. Here, $n_{d\sigma} = d_{\sigma}^{\dagger} d_{\sigma}$ represents the occupation number operator for dot electrons with spin σ. Because of the Coulomb interaction, there are two energy levels in the QD in the absence of magnetic fields. The singly and doubly occupied dot energies are ε_d and $2\varepsilon_d + U$, respectively. In order to be the QD-graphene system into the Kondo regime, the chemical potential in leads μ_α is located well above ε_d but also it is situated well below $\varepsilon_d + U$.

The last term in relation (3.1), $\mathcal{H}_{\text{T}}$, represents the tunneling Hamiltonian, and models the coupling of the energy levels in QD to the graphene electrodes,

$$\mathcal{H}_{\text{T}} = \sum_{\alpha,s,\sigma} \int_{-k_c}^{+k_c} dk \big\{ \mathcal{V}_{\alpha s\sigma}(k) c_{\alpha sk\sigma}^{\dagger} d_{\sigma} + [\mathcal{V}_{\alpha s\sigma}(k)]^* d_{\sigma}^{\dagger} c_{\alpha sk\sigma} \big\}, \tag{3.4}$$

where $\mathcal{V}_{\alpha s\sigma}(k)$ stands for the tunneling amplitude, which determines the strength of coupling between the graphene leads and the dot. Due to the fact that we consider an armchair configuration of monolayer graphene leads (see Fig. 3.1), in such arrangement the distance between the sublattices A and B of graphene layer and the QD is the same in real space within an adequate approximation. This results in a superposition of the K and K' valleys in the energy dispersion of armchair edges (for more details see Sect. 1.3). Thus, the tunneling amplitude can be approximated as valley-independent. Also, for simplicity, symmetrical QD-lead couplings and spin-independent tunneling amplitudes are considered throughout this chapter, $\mathcal{V}_{\alpha s\sigma}(k) = \mathcal{V}(k) = \tilde{V}\sqrt{|k|}$ where $\tilde{V} = V_0\sqrt{\Omega_0 \pi}/2\pi$ with V_0 standing for the tunneling strength and Ω_0 representing the graphene unit cell area.

3.2.2 *Current Formula*

To avoid any possible consequence coming from the symmetry properties of the QD-graphene system, we apply an asymmetric bias voltage to the graphene leads via the chemical potentials as $eV = \mu_L - \mu_R$ where μ_R is fixed. Mostly, in experimental measurements the asymmetric biasing is used [152]. The current flowing from the α graphene lead into the QD for electrons with spin σ is expressed as [10, 11, 13, 15, 16]:

$$I_{\alpha\sigma} = \frac{ie}{h}\int_{-D}^{+D} d\varepsilon\, \Gamma_\alpha(\varepsilon)\{G^{<}_{d\sigma}(\varepsilon) + f_\alpha(\varepsilon)[G^{r}_{d\sigma}(\varepsilon) - G^{a}_{d\sigma}(\varepsilon)]\}, \tag{3.5}$$

where $G^{<}_{d\sigma}(\varepsilon)$ is the lesser Green's function of the dot, while $G^{r(a)}_{d\sigma}(\varepsilon)$ refers to the retarded (advanced) Green's function of the QD. In addition, $\Gamma_\alpha(\varepsilon) = 2\pi\eta|\varepsilon|\theta(D - |\varepsilon|)$ stands for the coupling strength between the QD and α graphene lead. Here, $\eta = 2(\tilde{V}/\hbar v_F)^2$ is a dimensionless coupling parameter between the massless Dirac fermions in graphene and the localized electrons in QD. Using Kirchhoff's law, the total spin-dependent current can be symmetrized as $I_\sigma = (I_{L\sigma} - I_{R\sigma})/2$ [16], and one finds

$$I_\sigma = \frac{e}{\hbar}\int_{-D}^{+D} d\varepsilon \frac{\Gamma_L(\varepsilon)\Gamma_R(\varepsilon)}{\Gamma_L(\varepsilon) + \Gamma_R(\varepsilon)}\big[f_L(\varepsilon) - f_R(\varepsilon)\big]\rho_{d\sigma}(\varepsilon), \tag{3.6}$$

where $\rho_{d\sigma}(\varepsilon)$ is the DOS of QD for electrons with spin σ, which generally can depend on temperature T and bias voltage eV, respectively. $\rho_{d\sigma}(\varepsilon)$ is defined as

$$\rho_{d\sigma}(\varepsilon) = -\frac{1}{\pi}\mathrm{Im}G^{r}_{d\sigma}(\varepsilon). \tag{3.7}$$

The differential conductance for $\Gamma_\alpha(\varepsilon) = \Gamma(\varepsilon) = 2\pi\eta|\varepsilon|\theta(D - |\varepsilon|)$ reads [151]:

$$\frac{dI_\sigma}{dV} = \frac{2e^2}{h}\pi^2\eta\int_{-D}^{+D} d\varepsilon|\varepsilon|\theta(D - |\varepsilon|)\left\{\big[f_L(\varepsilon) - f_R(\varepsilon)\big]\frac{\partial\rho_{d\sigma}(\varepsilon)}{\partial(eV)} - \rho_{d\sigma}(\varepsilon)\frac{\partial f_L(\varepsilon)}{\partial\varepsilon}\right\}. \tag{3.8}$$

The total current and total differential conductance are expressed as the sum of spin-dependent currents and differential conductances as $I = \sum_\sigma I_\sigma$ and $dI/dV = \sum_\sigma dI_\sigma/dV$, respectively.

3.2.3 *Green's Function of the Quantum Dot*

As we have seen in Sect. 3.2.2, the transport properties of the QD-graphene system can be explored in the knowledge of the retarded Green's function of the dot $G^{r}_{d\sigma}(\varepsilon)$ which determines the DOS of QD as well as the current and differential conductance through the dot. To calculate $G^{r}_{d\sigma}(\varepsilon)$, we use the EOM technique [25, 39, 41, 45, 46, 100, 151] combined with the broadly applied Lacroix's decoupling scheme [41]. The higher-order correlation functions are determined within the framework of the Meir approximation [10]. The detailed computations for determining the retarded Green's function are presented in the Appendix of this chapter. Thus, after straightforward calculations, the retarded Green's function of the QD, defined by $G^{r}_{d\sigma}(\varepsilon) = \langle\langle d_\sigma|d^\dagger_\sigma\rangle\rangle^{r}_{\varepsilon}$, is obtained as

$$G_{d\sigma}^r(\varepsilon) = \frac{1-\langle n_{d\bar{\sigma}}\rangle}{\varepsilon-\varepsilon_{d\sigma}-\Sigma_0^r(\varepsilon)+U\frac{\Sigma_{3\sigma}(\varepsilon)+\Sigma_{4\sigma}(\varepsilon)}{\varepsilon-\varepsilon_{d\sigma}-U-\Sigma_0^r(\varepsilon)-\Sigma_{1\sigma}(\varepsilon)-\Sigma_{2\sigma}(\varepsilon)}} + \frac{\langle n_{d\bar{\sigma}}\rangle}{\varepsilon-\varepsilon_{d\sigma}-\Sigma_0^r(\varepsilon)-U-U\frac{\Sigma_{1\sigma}(\varepsilon)+\Sigma_{2\sigma}(\varepsilon)-\Sigma_{3\sigma}(\varepsilon)-\Sigma_{4\sigma}(\varepsilon)}{\varepsilon-\varepsilon_{d\sigma}-\Sigma_0^r(\varepsilon)-\Sigma_{1\sigma}(\varepsilon)-\Sigma_{2\sigma}(\varepsilon)}}, \tag{3.9}$$

where $\Sigma_0^r(\varepsilon)$ represents the spin-independent noninteracting tunneling self-energy,

$$\Sigma_0^r(\varepsilon) = \sum_{\alpha,s}\int_{-k_c}^{+k_c} dk\frac{\mathcal{V}(k)^2}{\varepsilon^+-\varepsilon_k} = -2\eta\left(\varepsilon\ln\frac{|D^2-\varepsilon^2|}{\varepsilon^2}+i\pi|\varepsilon|\theta(D-|\varepsilon|)\right). \tag{3.10}$$

The spin-dependent self-energies are defined as

$$\Sigma_{i\sigma}(\varepsilon) = \sum_{\alpha,s}\int_{-k_c}^{+k_c} dk\frac{\mathcal{V}(k)^2}{\Omega_{k\sigma}^{(i)}(\varepsilon)} = -2\eta\left(\omega_{i\sigma}\ln\frac{|D^2-\omega_{i\sigma}^2|}{\omega_{i\sigma}^2}+i\pi|\omega_{i\sigma}|\theta(D-|\omega_{i\sigma}|)\right), \tag{3.11}$$

with $i=1,2$ and $\Omega_{k\sigma}^{(1)}(\varepsilon)\equiv\varepsilon-\varepsilon_k+(\varepsilon_{d\bar{\sigma}}-\varepsilon_{d\sigma})+i0^+$, $\Omega_{k\sigma}^{(2)}(\varepsilon)\equiv\varepsilon+\varepsilon_k-(\varepsilon_{d\bar{\sigma}}+\varepsilon_{d\sigma})-U+i0^+$, $\omega_{1\sigma}=\varepsilon-\sigma\Delta\varepsilon_d$ and $\omega_{2\sigma}=\varepsilon-2\varepsilon_d-U$. Note that 0^+ is a positive infinitesimal. We also have

$$\Sigma_{3\sigma}(\varepsilon) = \sum_{\substack{\alpha,s\\ \beta,r}}\iint_{-k_c}^{+k_c} dkdq\mathcal{V}(k)\mathcal{V}(q)\frac{\langle c_{\beta rq\bar{\sigma}}^\dagger c_{\alpha sk\bar{\sigma}}\rangle}{\Omega_{k\sigma}^{(1)}} \approx \sum_{\alpha,s}\int_{-k_c}^{+k_c} dk\frac{\mathcal{V}(k)^2}{\Omega_{k\sigma}^{(1)}(\varepsilon)}f_\alpha(\varepsilon_k), \tag{3.12}$$

$$\Sigma_{4\sigma}(\varepsilon) = \sum_{\substack{\alpha,s\\ \beta,r}}\iint_{-k_c}^{+k_c} dkdq\mathcal{V}(k)\mathcal{V}(q)\frac{\langle c_{\alpha sk\bar{\sigma}}^\dagger c_{\beta rq\bar{\sigma}}\rangle}{\Omega_{k\sigma}^{(2)}} \approx \sum_{\alpha,s}\int_{-k_c}^{+k_c} dk\frac{\mathcal{V}(k)^2}{\Omega_{k\sigma}^{(2)}(\varepsilon)}f_\alpha(\varepsilon_k). \tag{3.13}$$

Note that by expressing Eq. (3.10) as $\Sigma_0^r(\varepsilon)=\sum_\alpha[\Lambda_\alpha(\varepsilon)-\frac{i}{2}\Gamma_\alpha(\varepsilon)]$ [16], the α graphene lead-QD coupling strength $\Gamma_\alpha(\varepsilon)$ can be derived to agree with the generally employed form in the literature, and assuming a symmetrically coupled QD system, i.e., $\Gamma_L(\varepsilon)=\Gamma_R(\varepsilon)$, one obtains $\Gamma_\alpha(\varepsilon)=2\pi\eta\rho_{\text{gr}}(\varepsilon)$ where $\rho_{\text{gr}}(\varepsilon)=|\varepsilon|\theta(D-|\varepsilon|)$ represents the DOS for graphene electrodes [100, 103]. As we have mentioned above, the Meir approximation [10] has been employed when performing the integral calculations appearing in $\Sigma_{3\sigma}(\varepsilon)$ and $\Sigma_{4\sigma}(\varepsilon)$, given by Eqs. (3.12) and (3.13), respectively. This approximation is based on non-self-consistent calculations of the average values of mixing operators. Namely, the terms coming from different operators are neglected, i.e., $\langle d_{\bar{\sigma}}^\dagger c_{\alpha sk\bar{\sigma}}\rangle=\langle c_{\alpha sk\bar{\sigma}}^\dagger d_{\bar{\sigma}}\rangle\approx 0$, and those corresponding to the same type of operators are approximated with the Fermi-Dirac function as $\langle c_{\beta rq\bar{\sigma}}^\dagger c_{\alpha sk\bar{\sigma}}\rangle=\langle c_{\alpha sk\bar{\sigma}}^\dagger c_{\beta rq\bar{\sigma}}\rangle\approx f_\alpha(\varepsilon_k)\delta_{\alpha\beta}\delta_{sr}\delta(k-q)$. We note here that the integral appearing in $\Sigma_{3\sigma}(\varepsilon)$, without magnetic field ($\Delta\varepsilon_d=0$), was analytically calculated in Ref. [100]. In our work [151], we have developed a simpler method for integral calculation with improved accuracy. A comparison of a special case of the technique discussed in Ref. [100] with our method, including numerical results, is presented in our article [151]. Our findings show a closer agreement with the numerical results compared to those obtained using the technique in Ref. [100]. The reason can be attributed to the fact that the integrand appearing in $\Sigma_{3\sigma}(\varepsilon)$ is not holomorphic, which limits the applicability of the contour integral technique used in Ref. [100]. All the self-energies introduced above, $\Sigma_{3\sigma}(\varepsilon)$ and $\Sigma_{4\sigma}(\varepsilon)$, respectively, are analytically evaluated in our published paper [151].

The average occupation number of electrons with spin σ in the dot is determined self-consistently by employing the relation [25, 31, 153]:

$$\langle n_{d\sigma}\rangle = \frac{1}{2\pi i}\int d\varepsilon\, G^{<}_{d\sigma}(\varepsilon), \tag{3.14}$$

where the lesser Green's function of the dot $G^{<}_{d\sigma}(\varepsilon)$ is calculated within the Keldysh equation

$$G^{<}_{d\sigma}(\varepsilon) = G^{r}_{d\sigma}(\varepsilon)\Sigma^{<}_{\sigma}(\varepsilon)G^{a}_{d\sigma}(\varepsilon), \tag{3.15}$$

where $G^{r(a)}_{d\sigma}(\varepsilon)$ and $\Sigma^{<}_{\sigma}(\varepsilon)$ are the retarded (advanced) Green's function of the dot and lesser self-energy, respectively. For interacting QDs, the retarded Green's function implicitly includes Coulomb interactions, and the self-energies must also be calculated by accounting for electron-electron interactions. Following Refs. [153–155], the lesser (greater) self-energy $\Sigma^{<(>)}_{\sigma}(\varepsilon)$ can be written as a product of a noninteracting term $\Sigma^{<(>)}_{0\sigma}(\varepsilon)$ and a function A, as $\Sigma^{<(>)}_{\sigma}(\varepsilon) = A\Sigma^{<(>)}_{0\sigma}(\varepsilon)$, where A is calculated from the relations $\Sigma^{>}_{\sigma}(\varepsilon) - \Sigma^{<}_{\sigma}(\varepsilon) = \Sigma^{r}_{\sigma}(\varepsilon) - \Sigma^{a}_{\sigma}(\varepsilon)$ and $\Sigma^{>}_{0\sigma}(\varepsilon) - \Sigma^{<}_{0\sigma}(\varepsilon) = \Sigma^{r}_{0\sigma}(\varepsilon) - \Sigma^{a}_{0\sigma}(\varepsilon)$, respectively. Here, $\Sigma^{r(a)}_{(0)\sigma}(\varepsilon)$ stands for the corresponding retarded (advanced) self-energy. For simplicity, we neglect the influence of Coulomb interaction U on the lesser self-energy, $\Sigma^{<}_{\sigma}(\varepsilon) \approx \Sigma^{<}_{0\sigma}(\varepsilon)$, assuming thus the corrections coming from the electron-electron interactions to the dot-lead couplings being negligible [23]. The noninteracting lesser self-energy reads

$$\Sigma^{<}_{0\sigma}(\varepsilon) = i\sum_{\alpha}\Gamma_{\alpha}(\varepsilon) f_{\alpha}(\varepsilon). \tag{3.16}$$

Using Ref. [16], and inserting Eq. (3.16) into Eq. (3.15), one finds

$$G^{<}_{d\sigma}(\varepsilon) = -2i\frac{\sum_{\alpha}\Gamma_{\alpha}(\varepsilon) f_{\alpha}(\varepsilon)}{\sum_{\alpha}\Gamma_{\alpha}(\varepsilon)}\mathrm{Im}G^{r}_{d\sigma}(\varepsilon), \tag{3.17}$$

which is reported also in Refs. [30, 153, 156].

3.2.4 Determination of a Kondo Temperature Formula

As we have discussed in Sect. 1.1.2, the Kondo temperature T_K is an energy scale which characterizes the system. We used the method applied in Refs. [28, 45, 157] in order to determine T_K. The non-self-consistent solutions of self-energies $\Sigma_{3\sigma}(\varepsilon)$ and $\Sigma_{4\sigma}(\varepsilon)$ are quantitatively correct at higher temperatures where the condition $T > T_K$ is fulfilled. At temperatures lower than the Kondo temperature, $T \leq T_K$, the obtained results, such as the estimated Kondo temperature and the lineshape of Kondo resonances, are only qualitatively correct [10, 31]. Note that the Kondo temperature is underestimated within the Lacroix approximation even for QDs connected to metallic leads while at the particle-hole symmetric point predicts a vanishing Kondo effect [25]. In contrast to all their limitations, the employed methods allow us to analyze the behavior of Kondo temperature and other quantities of our system [13, 15]. In the following, we deduce an estimated formula for the Kondo temperature T_K for arbitrary values of the Coulomb interaction U. Due to the fact that the self-energies $\Sigma_{3\sigma}(\varepsilon)$ and $\Sigma_{4\sigma}(\varepsilon)$ have different analytical solutions as a function of the sign of chemical potential μ_α (for details see our work [151]), the T_K will have different approximated solutions. Mathematically, the Kondo temperature is determined from the denominator of dot retarded Green's function. Thus, T_K is the temperature at which the value of the real part of the denominator of the retarded Green's function vanishes. In the case of strong Coulomb repulsion energy ($U \to \infty$), the first term of the retarded Green's function given by Eq. (3.9) holds due to its converging denominator. The second term of $G^r_{d\sigma}(\varepsilon)$ vanishes because of its diverging denominator. Thus, in order to estimate T_K, the first term of the retarded Green's function will be used. By introducing $\mathcal{N}(\varepsilon, T)$, the denominator of the first part of $G^r_{d\sigma}(\varepsilon)$ in equilibrium ($\mu_\alpha = \mu$) and without magnetic field ($\varepsilon_{d\sigma} = \varepsilon_d$), thus the solution of equation $\text{Re}\mathcal{N}(\varepsilon = \mu, T = T_K) = 0$ provides a formula for the Kondo temperature. In the following calculations, in order to avoid any band effect associated to the definition of D, we assume that the chemical potential is much smaller than the cutoff energy in graphene, i.e., $|\mu| \ll D$ [79] with $|\mu|/D \to 0$ and consider the low temperature limit. Thus, the following equation reads

$$\text{Re}\Sigma^{(\gamma)}_{3\sigma} + \text{Re}\Sigma^{(\gamma)}_{4\sigma} \approx -\left(c_1 + \frac{c_2^2}{c_1}\right)\left(\frac{\text{Im}\Sigma^{(\gamma)}_{3\sigma} + \text{Im}\Sigma^{(\gamma)}_{4\sigma}}{c_2^{-1}c_1^2 + c_2} + \frac{\varepsilon_d + \text{Re}\Sigma^r_0}{U}\right), \tag{3.18}$$

with notations [151]:

$$\begin{cases} c_1 \approx \varepsilon_d + U + \text{Re}\Sigma^r_0 + \text{Re}\Sigma_{1\sigma} + \text{Re}\Sigma_{2\sigma} \\ c_2 = \text{Im}\Sigma^r_0 + \text{Im}\Sigma_{1\sigma} + \text{Im}\Sigma_{2\sigma}, \end{cases} \tag{3.19}$$

where $\Sigma_{i\sigma} \equiv \Sigma_{i\sigma}(\varepsilon = \mu \to 0, T = T_K)$. To obtain approximate analytical results, we found that the following condition $2T_K < 2\varepsilon_d + U - \mu < 2D$ has to be fulfilled, which restricts the possibility of T_K's investigation at the particle-hole symmetry point ($2\varepsilon_d + U = 0$). Note that the Kondo temperature is the smallest energy scale in the system, satisfying $T_K \ll |\varepsilon_d|,\ \varepsilon_d + U,\ \Gamma_\alpha(\varepsilon),\ D$. As we have discussed, T_K depends on the sign of μ. The chemical potential can be modified by using various methods. We have two cases: (i) when $\mu > 0$, the graphene leads are electron-doped; (ii) when $\mu < 0$, it corresponds to hole doping of the graphene electrodes. In the (i) case when $\mu > 0$, using Eqs. (3.18)–(3.19), we obtain the following relation

$$T_K^{(+)} \propto \frac{|2\varepsilon_d + U - \mu|^3}{2e^2} \left| \frac{D^2 - (2\varepsilon_d + U)^2}{(2\varepsilon_d + U)^2} \right|^{\frac{2\varepsilon_d}{U}} \times \left| \frac{D + 2\varepsilon_d + U}{(2\varepsilon_d + U)^2} \right|^2 \exp\left\{ \frac{1}{\eta} \frac{\varepsilon_d(\varepsilon_d + U)}{U(2\varepsilon_d + U)} \right\}. \tag{3.20}$$

In the (ii) case when $\mu < 0$, the Kondo temperature reads

$$T_K^{(-)} \propto \frac{|2\varepsilon_d + U - \mu|^3}{2e^2|D + 2\varepsilon_d + U|^2} \left| \frac{D^2 - (2\varepsilon_d + U)^2}{(2\varepsilon_d + U)^2} \right|^{-\frac{2\varepsilon_d}{U}} \times \exp\left\{ -\frac{1}{\eta} \frac{\varepsilon_d(\varepsilon_d + U)}{U(2\varepsilon_d + U)} \right\}. \tag{3.21}$$

Note that in Eqs. (3.20) and (3.21) the symbol e stands for Euler's number. The properties of Lambert W function [158] have been used to derive $T_K^{(\gamma)}$, where $\gamma = +$ corresponds to $\mu > 0$ and $\gamma = -$ to $\mu < 0$, respectively.

The correctness of Eqs. (3.20)–(3.21) is verified by comparing our findings for graphene leads with the results of a system realized from a QD coupled to metallic electrodes, treated also within the Lacroix approximation [25, 45, 46]. They found, for finite U and metallic leads, that the Kondo temperature reads

$$T_K = (2\varepsilon_0 + U) \exp\left\{ \frac{2\pi\varepsilon_0(\varepsilon_0 + U)}{\Gamma U} \right\}, \tag{3.22}$$

where $\varepsilon_0 = \varepsilon_d - \mu$. It can be observed from the above relation that T_K does not depend on D for metallic leads, while $T_K^{(\gamma)}$ is also modulated by D for graphene electrodes because it determines the validity of the linear dispersion in graphene. We also see that T_K remains finite at the particle-hole symmetric point ($2\varepsilon_d + U = 0$) with $\mu \neq 0$ for normal leads, within the framework of Lacroix approximation. In our case, due to the introduced restriction for system parameters, we can only approach the particle-hole symmetric point and thus estimate $T_K^{(\gamma)}$ near the particle-hole symmetry point. To do this, we introduce a small quantity $\delta \equiv 2\varepsilon_d + U$ with $|\delta| \ll D$ in the vicinity of particle-hole symmetric point. Thus, when $\mu \equiv \mu^{(+)} > 0$, one obtains

$$T_K^{(+)} \propto \frac{|\delta - \mu^{(+)}|^3}{2e^2\delta^2} \exp\left\{ \frac{\delta^2 - U^2}{4\eta\delta U} \right\}, \tag{3.23}$$

and in the case of $\mu \equiv \mu^{(-)} < 0$, we have

$$T_K^{(-)} \propto \frac{|\delta - \mu^{(-)}|^3}{2e^2\delta^2} \exp\left\{ -\frac{\delta^2 - U^2}{4\eta\delta U} \right\}. \tag{3.24}$$

It seems from Eqs. (3.23) and (3.24) that $T_K^{(\gamma)}$ takes a non-zero value near the particle-hole symmetric point as in the case of metallic leads. The Kondo temperature T_K vanishes at $\mu = 0$ in particle-hole symmetry for QD-metallic electrodes structures. In the case of QD-graphene systems, we first consider that $\mu^{(+)} \to 0^+$ and $\delta \to 0^+$ as well as $U > |2\varepsilon_d|$ to be fulfilled, then from Eq. (3.23) reads that $T_K^{(+)} \approx 0$. Second, we assume that $\mu^{(-)} \to 0^-$ and $\delta \to 0^-$ with the condition $U < |2\varepsilon_d|$ and from Eq. (3.24) one obtains that $T_K^{(-)} \approx 0$. In addition, the vanishing Kondo temperature at the Dirac point in graphene-based systems has been reported in the literature [35, 104, 105, 122]. The NRG studies for a system based on a pseudogap Anderson Hamiltonian establish the absence of Kondo screening in particle-hole symmetry at $\mu = 0$. In particle-hole asymmetry, the Kondo screening manifests if the coupling constant exceeds a critical value [106]. Further NRG calculations show that the Kondo temperature is finite at $\mu = 0$ in particle-hole asymmetry case and it is symmetric in the chemical potential μ near the Dirac point [108]. We also note that Vojta and co-workers found within a combination of NRG and analytical renormalization-group calculations that T_K behaves differently depending on the sign of chemical potential μ in the limit of $U \to \infty$ [122]. Hence, if $\mu > 0$, they concluded that the Kondo temperature satisfies the law $T_K \propto |\mu|^x$ where $x \approx 2.6$ being an universal exponent. In the case when $\mu < 0$, they established that $T_K \propto \kappa|\mu|$ where κ stands for a universal prefactor. Therefore, an extreme asymmetrical behavior in T_K can be observed as the graphene is variously doped. Recently, Yanagisawa derived a Kondo temperature formula by employing Green's function theory [104]. It was established that T_K fulfills the law $T_K \propto |\mu|$ for small values of the chemical potential, being thus independent of the sign of μ. Furthermore, asymmetrical behavior of T_K regarding the position of μ, located above or below the Dirac point, has been reported in Ref. [103] for a system composed from depositing of a magnetic adatom on a graphene surface with a strong Coulomb interaction ($U \to \infty$).

As a conclusion, we found that the Kondo temperature vanishes at the Dirac point near the particle-hole symmetric point in the QD. In addition, away from the particle-hole symmetric point, the Kondo temperature remains finite even at the Dirac point which result is in agreement with NRG calculations in the finite U limit [108]. Based on previous studies, it can be established that the correct behavior of the Kondo temperature as a function of chemical potential is still not completely clarified even in the $U \to \infty$ limit for graphene-based systems. Additional analyses are needed to understand the behavior of Kondo temperature, also in the case of finite U.

Note here that a more correct estimation of the Kondo temperature needs self-consistent calculations rather then the non-self-consistent technique proposed by Meir et al. [10] for the determination of average values of mixing operators $\langle d_{\bar{\sigma}}^{\dagger} c_{\alpha sk\bar{\sigma}} \rangle$ and $\langle c_{\beta rq\bar{\sigma}}^{\dagger} c_{\alpha sk\bar{\sigma}} \rangle$ appearing in the retarded Green's function given by Eq. (3.9). At equilibrium ($f_\alpha(\varepsilon') = f(\varepsilon')$), they are calculated as [25, 41, 100, 152]

$$\langle d_{\bar{\sigma}}^{\dagger} c_{\alpha sk\bar{\sigma}} \rangle = -\frac{1}{\pi} \int_{-D}^{+D} d\varepsilon' f(\varepsilon') \mathrm{Im} \langle\langle c_{\alpha sk\bar{\sigma}} | d_{\bar{\sigma}}^{\dagger} \rangle\rangle_{\varepsilon'}^{r}, \tag{3.25}$$

$$\langle c^{\dagger}_{\beta rq\bar{\sigma}} c_{\alpha sk\bar{\sigma}} \rangle = -\frac{1}{\pi} \int_{-D}^{+D} d\varepsilon' f(\varepsilon') \mathrm{Im} \langle\langle c_{\alpha sk\bar{\sigma}} | c^{\dagger}_{\beta rq\bar{\sigma}} \rangle\rangle^{r}_{\varepsilon'}, \tag{3.26}$$

where the new correlation functions $\langle\langle c_{\alpha sk\bar{\sigma}} | d^{\dagger}_{\bar{\sigma}} \rangle\rangle^{r}_{\varepsilon'}$ and $\langle\langle c_{\alpha sk\bar{\sigma}} | c^{\dagger}_{\beta rq\bar{\sigma}} \rangle\rangle^{r}_{\varepsilon'}$ can be determined by using the EOM method. Note that, out-of-equilibrium, the expected values of mixing operators, $\langle d^{\dagger}_{\bar{\sigma}} c_{\alpha sk\bar{\sigma}} \rangle$ and $\langle c^{\dagger}_{\beta rq\bar{\sigma}} c_{\alpha sk\bar{\sigma}} \rangle$, are calculated using a modified fluctuation-dissipation theorem [30, 156]. On the other hand, to calculate the retarded Green's function $G^{r}_{d\sigma}(\varepsilon)$, instead of using the Lacroix approximation [41], a more accurate decoupling scheme [25] is needed. Within this method, the higher-order terms should be estimated based on self-consistent calculations. The resulting Green's function can be a promising candidate to the approximation of a more precise formula for the Kondo temperature. The decoupling scheme presented in Ref. [25] yields a non-vanishing Kondo temperature at the particle-hole symmetric point for QDs connected to normal leads. A potential direction for future research is the calculation of the retarded Green's function and Kondo temperature applying the methods outlined above. In addition, several methods have been developed for estimating the Kondo temperature. Some of the most commonly used approaches include perturbative renormalization group or "poor man's scaling" techniques [164–166], Bethe ansatz solutions [167], perturbation theory [168], conformal field theory [169], empirical fits to temperature-dependent linear conductance in QD experiments [42, 63], and extraction from the full width at half maximum of the Kondo resonance in differential conductance spectra [51].

3.3 Results and Discussion

In this chapter, we present the main numerical results performed for the QD-graphene nanodevice. To simplify the calculations, we use the energy cutoff D of graphene leads as the energy unit. The reduced Planck's and Boltzmann constants are set to unity, i.e., $\hbar = k_B = 1$. We choose $D \approx 7\,\mathrm{eV}$ as a reasonable value for graphene [38, 79, 80, 82]. We consider a GaAs QD with Landé factor $g = -0.44$ [70, 159, 160]. The Bohr magneton reads $\mu_B = 58\,\mu\mathrm{eV/T}$ [70]. The value of applied magnetic field is close to $B \approx 5\,\mathrm{T}$ according to the experimental measurements [70, 159, 160]. The resulting Zeeman splitting is $\Delta\varepsilon_d \approx 125\,\mu\mathrm{eV}$, i.e., $\Delta\varepsilon_d \approx 17 \cdot 10^{-6}\,D$, values close to those used in the following. The dot level is tuned by the voltage applied to the gate electrode [63]. The Coulomb energy U in the QD is chosen as $U \approx 480\,\mathrm{meV}$ that corresponds to $U = 0.069\,D$ which is a reasonable value for magnetic impurities in graphene [79, 80]. To estimate the dimensionless QD-graphene coupling parameter η, we consider a tunneling strength around $V_0 \approx 1\,\mathrm{eV}$ [82, 100] and assume that the area of unit cell in graphene is $\Omega_0 \approx 0.051\,\mathrm{nm}^2$ [161]. Consequently, the dimensionless coupling parameter reads $\eta \approx 0.0186$. In numerical calculations, values close to $\eta \approx 0.0186$ are used. We further note that the coupling strength η is a main system parameter which controls the transport regime. Specifically, when η

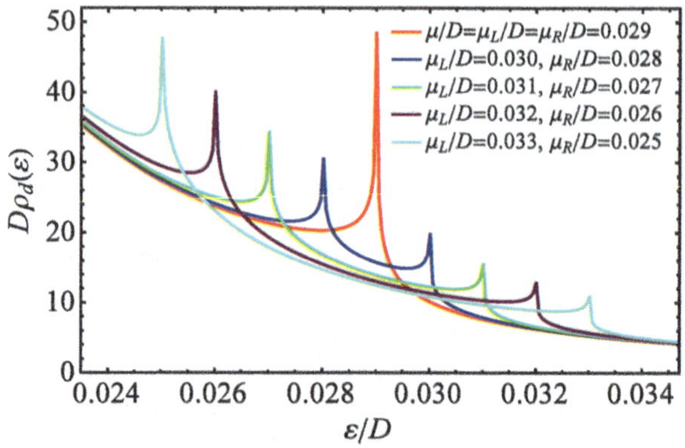

Fig. 3.2 The DOS of QD at different chemical potentials and temperature $T = 5 \cdot 10^{-6}\,D$ in the $U \to \infty$ limit without magnetic fields. The rest of parameters are the coupling $\eta = 0.02$ and QD energy $\varepsilon_d = -0.011\,D$. Reproduced from [151] under CC-BY 4.0 license. Copyright 2020 by the authors

coupling gradually increases, a transition in transport regimes from the Coulomb blockade to the Kondo regime occurs, as was found in the case of QDs connected to normal leads [23, 26, 27]. In the following discussions, we first explore the transport properties of the dot-graphene setup for a strongly interacting QD with $U \to \infty$ to better understand the phenomena which take place in such systems. After that, we generalize our investigation to finite U case.

In Fig. 3.2, we show the total DOS of QD when the values of chemical potentials in the leads are varied at zero magnetic field. The total DOS is determined as the sum of spin-dependent DOS, i.e., $\rho_d(\varepsilon) = \rho_{d\uparrow}(\varepsilon) + \rho_{d\downarrow}(\varepsilon)$. In equilibrium case when $\mu_L - \mu_R = eV = 0$, the Kondo resonance in DOS emerges at the equilibrium chemical potential $\mu_L = \mu_R = \mu$ which represents a resonant transmission through the dot [13]. An applied bias voltage between the graphene leads (at nonequilibrium, $\mu_L - \mu_R = eV \neq 0$) leads to the appearance of two peaks by splitting up the equilibrium Kondo resonance. The new peaks have smaller amplitudes and are located at the chemical potentials of left ($\varepsilon \approx \mu_L$) and right ($\varepsilon \approx \mu_R$) leads. The difference between chemical potentials, i.e., $\Delta\mu = \mu_L - \mu_R = eV$, determines the distance between the nonequilibrium Kondo peaks. The different chemical potentials values produce a dissipative transition of electrons by transferring them from the lead with higher chemical potential to the other one with lower chemical potential. This electron transfer results in a suppression of the peaks' amplitude and determines a relaxation time, denoted by τ_σ. The time τ_σ can be estimated within second-order perturbation calculations [13, 15, 31]. Here, we intuitively introduce it by replacing 0^+ with $\hbar/\tau_{\bar{\sigma}}$ in $\Omega^{(i)}_{k\sigma}(\varepsilon)$ of self-energies $\Sigma_{i\sigma}(\varepsilon)$. A small value for the relaxation time $\tau_{\bar{\sigma}}$ such as $\hbar/\tau_{\bar{\sigma}} \approx 10^{-7}$ [20] leads to a good agreement between the analytical and numerical results for self-energies $\Sigma_{i\sigma}(\varepsilon)$.

We plot, in Fig. 3.3, the nonequilibrium DOS of QD at different temperatures for different configurations of the system parameters without magnetic field. We observe

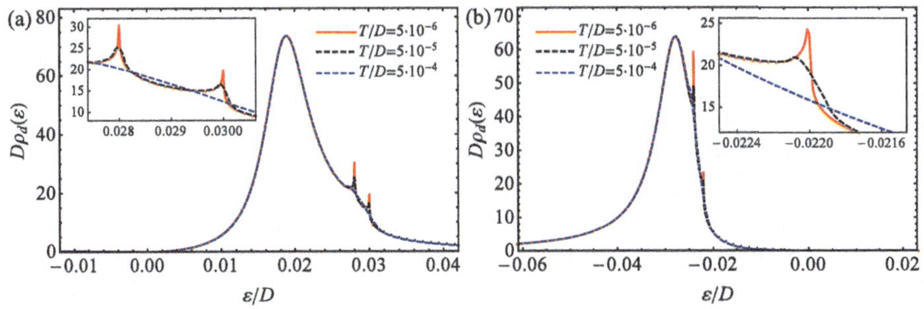

Fig. 3.3 The nonequilibrium DOS of QD at different temperatures T with $U \to \infty$ in the absence of magnetic field for the following choice of parameters: **a** $\mu_L = 0.028\, D$, $\mu_R = 0.030\, D$, $\eta = 0.02$ and $\varepsilon_d = -0.011\, D$. **b** $\mu_L = -0.024\, D$, $\mu_R = -0.022\, D$, $\eta = 0.015$ and $\varepsilon_d = -0.068\, D$. The inset in each Figure zooms in on a Kondo peak. Reproduced from [151] under CC-BY 4.0 license. Copyright 2020 by the authors

that the Kondo peaks emerge in the spectrum at the chemical potential of left ($\varepsilon \approx \mu_L$) and right ($\varepsilon \approx \mu_R$) leads. The Kondo resonances have a sharp lineshape at low temperatures and their magnitude decreases with the increase of temperature. At high temperatures, the Kondo peaks are totally washed out, in agreement with the results obtained previously for QDs connected to normal leads [13, 15]. Note that the lineshape of Kondo resonances is quantitatively not correct at temperatures close to the Kondo temperature, or at values lower than Kondo temperature within the approximations employed here. To approximate quantitatively correct the linshape of Kondo peaks, it is required to determine the mixed operators $\langle d^{\dagger}_{\bar{\sigma}} c_{\alpha s k \bar{\sigma}} \rangle$ and $\langle c^{\dagger}_{\beta r q \bar{\sigma}} c_{\alpha s k \bar{\sigma}} \rangle$ within a self-consistent calculation [25, 41, 100, 152] as pointed out in Chap. 3.2.4 when discussing the Kondo temperature. In addition, the DOS alongside the narrow Kondo resonances which strongly depend on temperature, contains a damped peak of large amplitude which shows up at $\varepsilon \approx \varepsilon_d + \mathrm{Re}\Sigma^r_0(\varepsilon) + \mathrm{Re}\Sigma_3(\varepsilon)$, where $\mathrm{Re}\Sigma_3(\varepsilon) = \mathrm{Re}\Sigma_{3\uparrow}(\varepsilon) = \mathrm{Re}\Sigma_{3\downarrow}(\varepsilon)$ in the absence of magnetic field. The broadened peak represents a resonant transmission at energy $\varepsilon \approx \tilde{\varepsilon}_d$. Here, $\tilde{\varepsilon}_d$ is a renormalized QD level which can be determined self-consistently by employing the relation $\tilde{\varepsilon}_d = \varepsilon_d + \mathrm{Re}\Sigma^r_0(\tilde{\varepsilon}_d) + \mathrm{Re}\Sigma_3(\tilde{\varepsilon}_d)$ [31].

Next, we turn our attention to the influence of applied magnetic field on transport properties of the QD-graphene system. We assume that the graphene electrodes are well enough separated from the QD in such a way that the magnetic field acts only on the dot without affecting the leads. This approach is a broadly employed approximation in the literature for QDs coupled to metallic electrodes [13, 18, 162]. The investigation of the spin-polarized transport properties of QDs attached to ferromagnetic leads is beyond of this book's scope. We plot in Fig. 3.4 the nonequilibrium DOS as a function of energy when a magnetic field is applied to the dot with different values of the Zeeman splitting $\Delta\varepsilon_d$, chemical potentials μ_α and QD-leads coupling η and $U \to \infty$ at a finite temperature T. The magnetic field results in a splitting of the Kondo peaks appearing initially at $\varepsilon = \mu_\alpha$ into two peaks with reduced amplitudes. The Kondo resonances are shifted by the Zeeman splitting from their nonmagnetic positions given by the chemical potentials to the higher energies for electrons with spin ↑ and to the lower energies for electrons with spin ↓. This behavior of the Kondo peaks is in agreement with the results of [13] for a QD coupled to normal leads in the presence of magnetic fields. Beside the sharp spin-dependent Kondo peaks, a broadened resonance shows up in the DOS for spin σ electrons. This large resonance represents a resonant tunneling of electrons with spin σ at the spin-dependent renormalized QD energy $\tilde{\varepsilon}_{d\sigma} = \varepsilon_{d\sigma} + \mathrm{Re}\Sigma_0^r(\tilde{\varepsilon}_{d\sigma}) + \mathrm{Re}\Sigma_{3\sigma}(\tilde{\varepsilon}_{d\sigma})$. It is important to note that the DOS of QD vanishes at $\varepsilon = 0$ independently of the spin orientation. Furthermore, the Kondo resonances do not emerge in the DOS when the chemical potential in leads is pinned at the Dirac point, i.e., $\mu_\alpha = 0$, in agreement with the results obtained for an adatom placed on a graphene surface with $U \to \infty$ [103]. This behavior of the DOS is a consequence of the linear dispersion of graphene electrodes.

The numerical results for the differential conductance dI/dV through the QD as a function of applied bias voltage eV are shown in Fig. 3.5a for different values of the temperature T in the absence of magnetic fields. The zero-bias peak develops in the differential conductance at low temperatures when the difference in chemical potentials meets the Zeeman splitting, i.e., $eV = \Delta\mu = \Delta\varepsilon_d = 0$, which represents a resonant transmission through the dot. The zero-bias peak of dI/dV plots is strongly temperature-dependent. At low temperatures, the Kondo peak has a sharper and narrow lineshape and with the grow in temperature, the amplitude of resonance reduces and broadens. In addition, the lineshape of the Kondo peak in dI/dV is strongly bias voltage eV dependent. Thus, the magnitude of the peak takes its maximum value at $eV = 0$ and it quickly drops when the bias voltage eV deviates from $eV = 0$. The changes in dI/dV with respect to bias voltage near $eV = 0$ can be approximated with the help of a second-degree polynomial function. The variation of Kondo peak with eV shows similar tendency with the findings of Świrkowicz et al. for a system

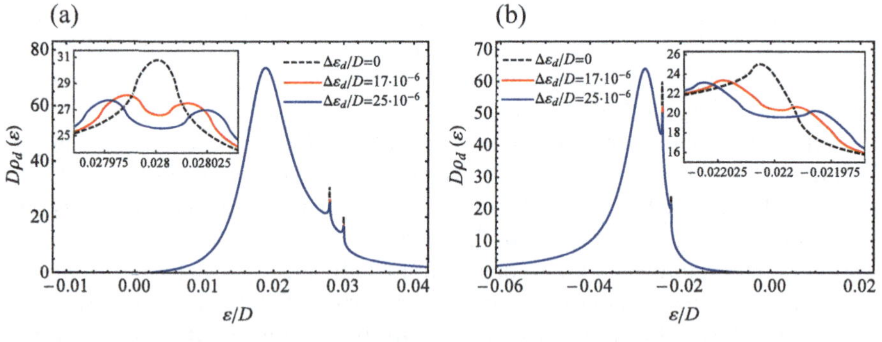

Fig. 3.4 The nonequilibrium DOS of QD at different Zeeman splitting $\Delta\varepsilon_d$ values and temperature $T = 5\cdot10^{-6}\,D$ with $U \to \infty$ for the following choice of parameters: **a** $\mu_L = 0.028\,D$, $\mu_R = 0.030\,D$, $\eta = 0.02$ and $\varepsilon_d = -0.011\,D$. **b** $\mu_L = -0.024\,D$, $\mu_R = -0.022\,D$, $\eta = 0.015$ and $\varepsilon_d = -0.068\,D$. The inset in each Figure zooms in on a Kondo peak. Reproduced from [151] under CC-BY 4.0 license. Copyright 2020 by the authors

realized from a QD attached to normal leads [20]. Figure 3.5b shows the numerical results for the differential conductance dI/dV against the bias voltage eV for different Zeeman splitting $\Delta\varepsilon_d$ values and at a fixed temperature $T = 5 \cdot 10^{-6}\, D$. We observe that the applied magnetic field on QD leads to the splitting of zero-bias peak into two resonances with reduced amplitudes, showing up at voltages $eV = \pm\Delta\varepsilon_d$. At small values of the Zeeman splitting, the split Kondo resonances become broadened due to their overlapping [18]. Our results show a similar behavior with those systems realized from QDs connected to metallic electrodes threaded by magnetic fields [13, 18]. The zero-bias peak splitting in differential conductance measurements was detected for cobalt adatoms placed on a graphene monolayer [124].

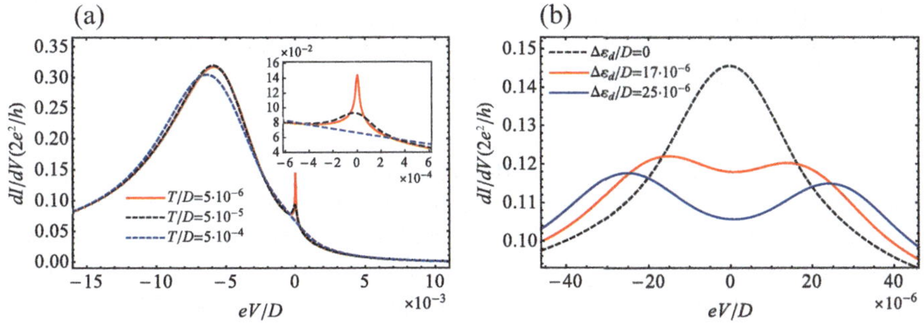

Fig. 3.5 **a** Differential conductance dI/dV against the bias voltage eV at different temperatures T and $\mu_R = -0.022\, D$ without magnetic fields. **b** Differential conductance dI/dV against the bias voltage eV at different Zeeman splitting $\Delta\varepsilon_d$ values and temperature $T = 5 \cdot 10^{-6}\, D$. The remaining system parameters are same as in Fig. 3.3b. The inset in **a** zoom in on the zero-bias peak. Reproduced from [151] under CC-BY 4.0 license. Copyright 2020 by the authors

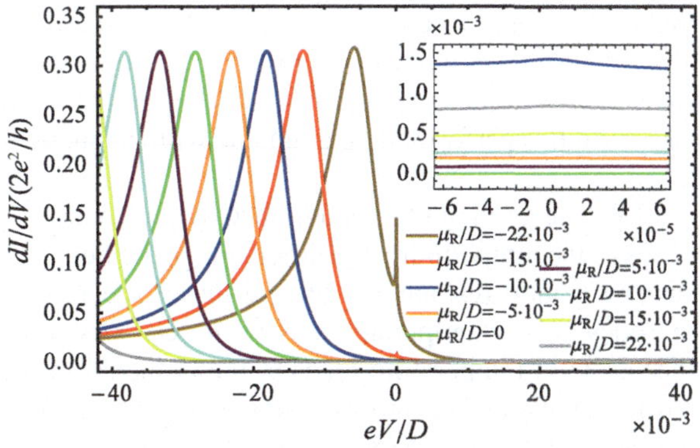

Fig. 3.6 Differential conductance dI/dV against the bias voltage eV at different chemical potential μ_R values and temperature $T = 5 \cdot 10^{-6}\, D$ without magnetic fields. The remaining system parameters are same as in Fig. 3.5a. The inset zoom in on the zero-bias peak. Reprinted from [151] under CC-BY 4.0 license. Copyright 2020 by the authors

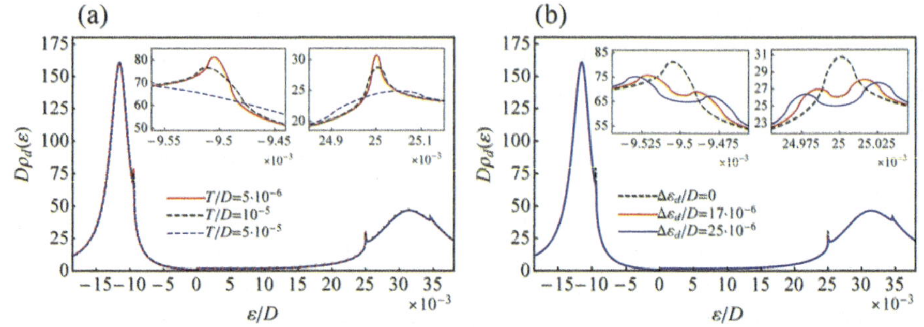

Fig. 3.7 **a** The nonequilibrium DOS of QD at different temperatures T without magnetic fields. **b** The nonequilibrium DOS of QD at different Zeeman splitting $\Delta\varepsilon_d$ values at temperature $T = 5 \cdot 10^{-6}\,D$. The rest of parameters are: $U = 0.069\,D$, $\mu_L = 25 \cdot 10^{-3}\,D$, $\mu_R = -9.5 \cdot 10^{-3}\,D$, $\eta = 0.02$ and $\varepsilon_d = -0.022\,D$. The insets zoom in on the Kondo peaks. Reproduced from [151] under CC-BY 4.0 license. Copyright 2020 by the authors

Figure 3.6 shows the variation of differential conductance dI/dV with the applied bias voltage eV at different values of the chemical potential in the right graphene lead, μ_R, at temperature $T = 5 \cdot 10^{-6}\,D$, without magnetic field ($\Delta\varepsilon_d = 0$). We observe that when μ_R matches the Dirac point, the zero-bias peak does not emerge in the dI/dV plots due to the linear dispersion of graphene electrodes. The zero-bias resonance shows up in the differential conductance only when the chemical potential is located above or below the Dirac point in graphene leads revealing the presence of Kondo effect [103]. We see that the zero-bias peak magnitude is strongly μ_R-dependent. In the case when μ_R is located below the Dirac point in graphene, then with further increase in $|\mu_R|$ the zero-bias peak becomes higher and dominant. We also observe that when μ_R is situated above the graphene Dirac point, the Kondo peak is much smaller compared to the $\mu_R < 0$ case. The discrepancy in the height of zero-bias peaks, when μ_R has different signs, can be explained by the presence of particle-hole asymmetry in the system. This tendency of the zero-bias peak is in agreement with other studies performed on graphene with magnetic impurities [100, 103, 122].

We plot, in Fig. 3.7a, the DOS of QD for a finite Coulomb energy $U = 0.069\,D$ at three different values of the temperature T, without magnetic field. It can be observed that sharp and narrow Kondo resonances emerge in the spectrum at the chemical potentials of electrodes at low temperatures as in the $U \to \infty$ limit (see Figs. 3.2 and 3.3). A further grow in temperature leads to the reduction of the amplitude of resonances and to their totally disappearance at high temperatures. In addition, beside the strongly temperature-dependent Kondo peaks, two broadened resonances show up in the DOS at $\varepsilon \approx \mathrm{Re}\mathcal{N}_1(\varepsilon)$ and $\varepsilon \approx \mathrm{Re}\mathcal{N}_2(\varepsilon)$ with large amplitudes where $\varepsilon - \mathcal{N}_1(\varepsilon)$ and $\varepsilon - \mathcal{N}_2(\varepsilon)$ denote the denominator of the first and second terms of retarded Green's function given by Eq. (3.9) at zero magnetic field. The broadened peak appearing at lower energy refers to a resonant transmission through the dot at $\varepsilon \approx \tilde{\varepsilon}_d$ and the other one, showing up at higher energy, corresponds to a transmission at $\varepsilon \approx$

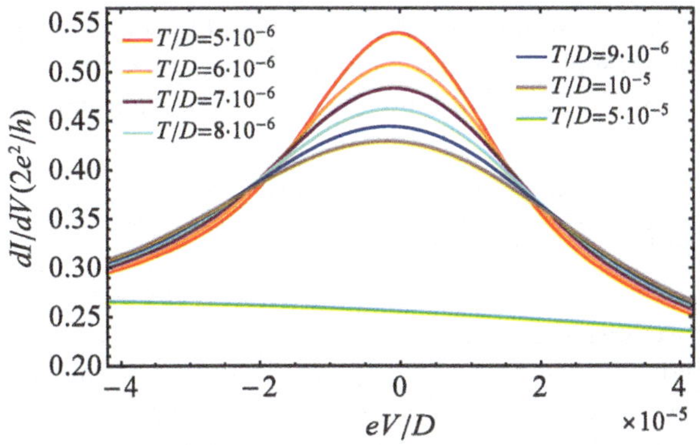

Fig. 3.8 Differential conductance dI/dV against the bias voltage eV with $\mu_R = -9.5 \cdot 10^{-3}\,D$ at different values of the temperature T for a finite U without magnetic fields. The remaining system parameters are same as in Fig. 3.7a. Reprinted from [151] under CC-BY 4.0 license. Copyright 2020 by the authors

$\tilde{\varepsilon}_d + \tilde{U}$. Here, $\tilde{\varepsilon}_d$ and $\tilde{U}$ are the renormalized dot level and Coulomb repulsion energy that need to be determined self-consistently by employing the method proposed by Van Roermund et al. [25]. Figure 3.7b illustrates the results of the total DOS with finite Coulomb energy $U = 0.069\,D$ at temperature $T = 5 \cdot 10^{-6}\,D$ in the presence of an external magnetic field. As in the strongly interacting QD case with $U \to \infty$ (see Fig. 3.4), the Kondo peaks split into two peaks due to the applied magnetic field on the QD. The new peaks with reduced amplitudes move by the Zeeman splitting $\Delta\varepsilon_d$ from thc chemical potentials (μ_L and μ_R), to right for electrons with spin $\uparrow$, and to the left for electrons with spin $\downarrow$. Two large broadened peaks develop in the DOS at $\varepsilon \approx \mathrm{Re}\mathcal{N}_{1\sigma}(\varepsilon)$ and $\varepsilon \approx \mathrm{Re}\mathcal{N}_{2\sigma}(\varepsilon)$ for electrons with spin σ, corresponding to transport within the Coulomb-blockade regime when a magnetic field is applied to the dot. In the same way, we can estimate the values of the spin-dependent renormalized dot level $\tilde{\varepsilon}_{d\sigma}$ and renormalized Coulomb energy $\tilde{U}$.

In Fig. 3.8, we plot the total differential conductance dI/dV as a function of applied bias voltage eV for a finite Coulomb interaction $U = 0.069\,D$, at different values of the temperature T in the absence of a magnetic field. As in the $U \to \infty$ limit (see Fig. 3.5), the differential conductance's plot presents a zero-bias peak when the bias voltage eV matches the Zeeman splitting ($eV = \Delta\varepsilon_d = 0$) at low temperatures. The lineshape of the zero-bias peak is strongly temperature-dependent. Furthermore, the Kondo resonance in differential conductance is totally smeared at high temperatures. As the temperature decreases, the Kondo peak begins to manifest in the spectrum of dI/dV. Further decrease in T leads to the narrowing of the zero-bias differential conductance peak due to the sharp Kondo resonance in the DOS of QD that shows up at low temperatures. Therefore, near $eV \approx 0$, the lineshape of Kondo peak is strongly eV-dependent, i.e., its magnitude significantly reduces with increasing $|eV|$. As in the strongly interacting QD case ($U \to \infty$), the variation of dI/dV curve with eV around $eV = 0$ can be approximated with the help of a second-degree polynomial function at small values of the bias voltage $|eV|$. We note again that the variation of Kondo resonance curve with bias voltage presents similar tendency with the findings of [20] for a QD connected to normal leads.

3.4 Summary

In this chapter, we have theoretically investigated the nonequilibrium Kondo effect in a QD attached to two monolayer graphene leads in an armchair configuration. The QD-graphene system is modeled within a pseudogap Anderson Hamiltonian in the presence of a magnetic field. We have determined the retarded Green's function of the dot for arbitrary values of the Coulomb interaction by employing the EOM technique [25, 39, 41, 45, 46, 100, 151] within the Lacroix decoupling scheme [41] combined with the Meir approximation [10]. The retarded Green's function determines the transport properties of the system, such as the DOS of QD, current and differential conductance through the dot. We have estimated formulas for the Kondo temperature when the graphene leads are electron or hole doped for finite values of the Coulomb interaction. It has been found that the Kondo temperature vanishes near the particle-hole symmetry point when the chemical potential of graphene leads is tuned to the Dirac point. In the particle-hole asymmetry case, independent of the doping of graphene leads, the Kondo temperature holds, showing a finite value. Such behavior of Kondo temperature is in agreement with results based on other methods in the literature. The transport characteristics of the QD system have been explored in detail for both a strongly interacting QD and a QD with finite Coulomb interaction, with and without magnetic fields, at finite temperatures. The Coulomb interaction has the usual influence on transport properties as in the case of QDs coupled to normal leads. The DOS of QD disappears at zero-energy due to the linear energy dispersion of graphene leads. The Kondo peaks show up in QD DOS at chemical potentials of electrodes without magnetic fields. The presence of a magnetic field leads to the splitting of the Kondo resonances in the DOS. Accordingly, the zero-bias peak in differential conductance does not manifest itself when the chemical potential in the leads matches the Dirac point in graphene. The finite magnetic field results in a splitting of the zero-bias peak in the differential conductance. The proposed QD-graphene system illustrated in Fig. 3.1 can be implemented experimentally, allowing the verification of our findings, which can significantly contribute to the advancement of the knowledge in graphene-based nanoelectronics. Determining the Kondo temperature within a more accurate self-consistent calculation of the average values of mixed operators, rather than applying the Meir approximation [10] in the self-energies of the retarded Green's function, can play a key aspect of future research. By employing a more precise decoupling scheme [25], extending beyond the Lacroix approach [41], for the calculation of the retarded Green's function, could be a potential subject of further investigations.

Appendix

In this appendix, based on our paper [151], we determine the retarded Green's function of the dot, given in Eq. (3.9), using the EOM method discussed at the beginning of Appendix of second chapter. We define the dot retarded Green's function as $G^r_{d\sigma}(\varepsilon) = \langle\langle d_\sigma | d^\dagger_\sigma \rangle\rangle^r_\varepsilon$ when replacing $A(B)$ in $\langle\langle A|B\rangle\rangle^r_\varepsilon$ with $d_\sigma (d^\dagger_\sigma)$. Using relations (3.1)–(3.4), and (2.36)–(2.37), the EOM for $G^r_{d\sigma}(\varepsilon)$ is calculated as follows:

$$\varepsilon^+ \langle\langle d_\sigma | d^\dagger_\sigma \rangle\rangle^r_\varepsilon + \langle\langle [\mathcal{H}, d_\sigma] | d^\dagger_\sigma \rangle\rangle^r_\varepsilon = \langle\langle \{d_\sigma, d^\dagger_\sigma\} \rangle\rangle, \tag{3.27}$$

where $\{d_\sigma, d^\dagger_\sigma\} = 1$ and

$$[\mathcal{H}, d_\sigma] = [\mathcal{H}_{\text{Leads}}, d_\sigma] + [\mathcal{H}_{\text{QD}}, d_\sigma] + [\mathcal{H}_{\text{T}}, d_\sigma], \tag{3.28}$$

with

$$\begin{aligned}
[\mathcal{H}_{\text{Leads}}, d_\sigma] &= \sum_{\alpha,s,\eta} \int_{-k_c}^{+k_c} dk\, \varepsilon_k \left[c^\dagger_{\alpha s k \eta} c_{\alpha s k \eta}, d_\sigma\right] \\
&= -\sum_{\alpha,s,\eta} \int_{-k_c}^{+k_c} dk\, \varepsilon_k \Big(\underbrace{\{d_\sigma, c^\dagger_{\alpha s k \eta}\}}_{=0} c_{\alpha s k \eta} - c^\dagger_{\alpha s k \eta} \underbrace{\{d_\sigma, c_{\alpha s k \eta}\}}_{=0} \Big) = 0,
\end{aligned} \tag{3.29}$$

$$\begin{aligned}
[\mathcal{H}_{\text{QD}}, d_\sigma] &= \sum_\eta \varepsilon_{d\eta} [d^\dagger_\eta d_\eta, d_\sigma] + U\left[n_{d\uparrow} n_{d\downarrow}, d_\sigma\right] = -\sum_\eta \varepsilon_{d\eta} \Big(\underbrace{\{d_\sigma, d^\dagger_\eta\}}_{=\delta_{\sigma\eta}} d_\eta \\
&- d^\dagger_\eta \underbrace{\{d_\sigma, d_\eta\}}_{=0} \Big) - U\Big([d_\sigma, d^\dagger_\uparrow d_\uparrow] n_{d\downarrow} + n_{d\uparrow} [d_\sigma, d^\dagger_\downarrow d_\downarrow] \Big) = -\varepsilon_{d\sigma} d_\sigma \\
&- U\Big(\underbrace{\{d_\sigma, d^\dagger_\uparrow\}}_{=\delta_{\sigma\uparrow}} d_\uparrow - d^\dagger_\uparrow \underbrace{\{d_\sigma, d_\uparrow\}}_{=0} \Big) n_{d\downarrow} - U n_{d\uparrow} \Big(\underbrace{\{d_\sigma, d^\dagger_\downarrow\}}_{=\delta_{\sigma\downarrow}} d_\downarrow - d^\dagger_\downarrow \underbrace{\{d_\sigma, d_\downarrow\}}_{=0} \Big) \\
&= -\varepsilon_{d\sigma} d_\sigma - U d_\sigma n_{d\bar{\sigma}},
\end{aligned} \tag{3.30}$$

$$\begin{aligned}
[\mathcal{H}_{\text{T}}, d_\sigma] &= \sum_{\alpha,s,\eta} \int_{-k_c}^{+k_c} dk \Big(\mathcal{V}_{\alpha s \eta}(k) [c^\dagger_{\alpha s k \eta} d_\eta, d_\sigma] + [\mathcal{V}_{\alpha s \eta}(k)]^* [d^\dagger_\eta c_{\alpha s k \eta}, d_\sigma] \Big) \\
&= -\sum_{\alpha,s,\eta} \int_{-k_c}^{+k_c} dk \Big[\mathcal{V}_{\alpha s \eta}(k) \Big(\underbrace{\{d_\sigma, c^\dagger_{\alpha s k \eta}\}}_{=0} d_\eta - c^\dagger_{\alpha s k \eta} \underbrace{\{d_\sigma, d_\eta\}}_{=0} \Big)
\end{aligned}$$

$$+ [\mathcal{V}_{\alpha s\eta}(k)]^*\Big(\underbrace{\{d_\sigma, d_\eta^\dagger\}}_{=\delta_{\sigma\eta}} c_{\alpha sk\eta} - d_\eta^\dagger \underbrace{\{d_\sigma, c_{\alpha sk\eta}\}}_{=0}\Big)\Big]$$

$$= -\sum_{\alpha,s}\int_{-k_c}^{+k_c} dk\,[\mathcal{V}_{\alpha s\sigma}(k)]^*\, c_{\alpha sk\sigma}. \tag{3.31}$$

Substituting relations from (3.29)–(3.31) into Eqs. (3.28), (3.27) reads

$$(\varepsilon^+ - \varepsilon_{d\sigma})\langle\langle d_\sigma | d_\sigma^\dagger\rangle\rangle_\varepsilon^r = 1 + \sum_{\alpha,s}\int_{-k_c}^{+k_c} dk\,[\mathcal{V}_{\alpha s\sigma}(k)]^* \langle\langle c_{\alpha sk\sigma} | d_\sigma^\dagger\rangle\rangle_\varepsilon^r + U\langle\langle d_\sigma n_{d\bar{\sigma}} | d_\sigma^\dagger\rangle\rangle_\varepsilon^r. \tag{3.32}$$

The equation for the term $\langle\langle c_{\alpha sk\sigma} | d_\sigma^\dagger\rangle\rangle_\varepsilon^r$ reads

$$\varepsilon^+\langle\langle c_{\alpha sk\sigma} | d_\sigma^\dagger\rangle\rangle_\varepsilon^r + \langle\langle [\mathcal{H}, c_{\alpha sk\sigma}] | d_\sigma^\dagger\rangle\rangle_\varepsilon^r = \langle\{c_{\alpha sk\sigma}, d_\sigma^\dagger\}\rangle, \tag{3.33}$$

where $\{c_{\alpha sk\sigma}, d_\sigma^\dagger\} = 0$ and using Eqs. (2.37) and (3.1), one finds

$$[\mathcal{H}, c_{\alpha sk\sigma}] = [\mathcal{H}_{\text{Leads}}, c_{\alpha sk\sigma}] + [\mathcal{H}_{\text{QD}}, c_{\alpha sk\sigma}] + [\mathcal{H}_{\text{T}}, c_{\alpha sk\sigma}], \tag{3.34}$$

with

$$\begin{aligned}[\mathcal{H}_{\text{Leads}}, c_{\alpha sk\sigma}] &= \sum_{\beta,r,\eta}\int_{-k_c}^{+k_c} dq\,\varepsilon_q\left[c_{\beta rq\eta}^\dagger c_{\beta rq\eta}, c_{\alpha sk\sigma}\right] \\ &= -\sum_{\beta,r,\eta}\int_{-k_c}^{+k_c} dq\,\varepsilon_q\Big(\underbrace{\{c_{\alpha sk\sigma}, c_{\beta rq\eta}^\dagger\}}_{=\delta_{\alpha\beta}\,\delta_{sr}\,\delta_{\sigma\eta}\,\delta(k-q)} c_{\beta rq\eta} - c_{\beta rq\eta}^\dagger\underbrace{\{c_{\alpha sk\sigma}, c_{\beta rq\eta}\}}_{=0}\Big) \\ &= -\varepsilon_k\, c_{\alpha sk\sigma},\end{aligned} \tag{3.35}$$

$$\begin{aligned}[\mathcal{H}_{\text{QD}}, c_{\alpha sk\sigma}] &= \sum_\eta \varepsilon_{d\eta}\left[d_\eta^\dagger d_\eta, c_{\alpha sk\sigma}\right] + U\left[n_{d\uparrow}n_{d\downarrow}, c_{\alpha sk\sigma}\right] = -\sum_\eta \varepsilon_{d\eta}\Big(\underbrace{\{c_{\alpha sk\sigma}, d_\eta^\dagger\}}_{=0} d_\eta \\ &\quad - d_\eta^\dagger\underbrace{\{c_{\alpha sk\sigma}, d_\eta\}}_{=0}\Big) - U\Big([c_{\alpha sk\sigma}, d_\uparrow^\dagger d_\uparrow] n_{d\downarrow} + n_{d\uparrow}[c_{\alpha sk\sigma}, d_\downarrow^\dagger d_\downarrow]\Big) \\ &= -U\Big(\underbrace{\{c_{\alpha sk\sigma}, d_\uparrow^\dagger\}}_{=0} d_\uparrow - d_\uparrow^\dagger\underbrace{\{c_{\alpha sk\sigma}, d_\uparrow\}}_{=0}\Big) n_{d\downarrow} \\ &\quad - U n_{d\uparrow}\Big(\underbrace{\{c_{\alpha sk\sigma}, d_\downarrow^\dagger\}}_{=0} d_\downarrow - d_\downarrow^\dagger\underbrace{\{c_{\alpha sk\sigma}, d_\downarrow\}}_{=0}\Big) = 0,\end{aligned} \tag{3.36}$$

and

$$
\begin{aligned}
[\mathcal{H}_{\mathrm{T}}, c_{\alpha sk\sigma}] &= \sum_{\beta,r,\eta} \int_{-k_c}^{+k_c} dq \Big(\mathcal{V}_{\beta r\eta}(q) \big[c^{\dagger}_{\beta rq\eta} d_{\eta}, c_{\alpha sk\sigma} \big] + [\mathcal{V}_{\beta r\eta}(q)]^{*} \big[d^{\dagger}_{\eta} c_{\beta rq\eta}, c_{\alpha sk\sigma} \big] \Big) \\
&= -\sum_{\beta,r,\eta} \int_{-k_c}^{+k_c} dq \Big[\mathcal{V}_{\beta r\eta}(q) \big(\underbrace{\{c_{\alpha sk\sigma}, c^{\dagger}_{\beta rq\eta}\}}_{=\delta_{\alpha\beta}\,\delta_{sr}\,\delta_{\sigma\eta}\,\delta(k-q)} d_{\eta} - c^{\dagger}_{\beta rq\eta} \underbrace{\{c_{\alpha sk\sigma}, d_{\eta}\}}_{=0} \big) \\
&+ [\mathcal{V}_{\beta r\eta}(q)]^{*} \big(\underbrace{\{c_{\alpha sk\sigma}, d^{\dagger}_{\eta}\}}_{=0} c_{\beta rq\eta} - d^{\dagger}_{\eta} \underbrace{\{c_{\alpha sk\sigma}, c_{\beta rq\eta}\}}_{=0} \big) \Big] \\
&= -\mathcal{V}_{\alpha s\sigma}(k)\, d_{\sigma}.
\end{aligned}
\tag{3.37}
$$

Substituting relations from (3.35)–(3.37) into Eqs. (3.34) and (3.33), we obtain

$$
\langle\langle c_{\alpha sk\sigma} | d^{\dagger}_{\sigma} \rangle\rangle^{r}_{\varepsilon} = \frac{\mathcal{V}_{\alpha s\sigma}(k)}{\varepsilon^{+} - \varepsilon_{k}} \langle\langle d_{\sigma} | d^{\dagger}_{\sigma} \rangle\rangle^{r}_{\varepsilon}. \tag{3.38}
$$

Following Refs. [39, 93, 100], and using Eq. (3.38) with (3.32), we can define a $\Sigma^{r}_{0}(\varepsilon)$ spin-independent noninteracting tunneling self-energy as

$$
\Sigma^{r}_{0}(\varepsilon) = \sum_{\alpha,s} \int_{-k_c}^{+k_c} dk \frac{\left|\mathcal{V}_{\alpha s\sigma}(k)\right|^{2}}{\varepsilon^{+} - \varepsilon_{k}} = -\sum_{\alpha} \eta_{\alpha} \left(\varepsilon \ln \left| \frac{D^{2} - \varepsilon^{2}}{\varepsilon^{2}} \right| + i\pi|\varepsilon|\theta(D - |\varepsilon|) \right), \tag{3.39}
$$

where we used the notation $\mathcal{V}_{\alpha s\sigma}(k) = \tilde{V}_{\alpha}\sqrt{|k|}$ with $\tilde{V}_{\alpha} = V_{\alpha}\sqrt{\Omega_{0}\pi}/2\pi$ and introduced a dimensionless coupling parameter as $\eta_{\alpha} = 2(\tilde{V}_{\alpha}/\hbar v_{F})^{2}$ with cutoff energy $D = \hbar v_{F} k_{c}$. Note that, for a symmetrically coupled QD system ($\eta_{\alpha} = \eta$), Eq. (3.39) reduces to the relation (3.10). Substituting Eq. (3.38) into Eq. (3.32), with Eq. (3.39), we find

$$
(\varepsilon^{+} - \varepsilon_{d\sigma} - \Sigma^{r}_{0}(\varepsilon)) \langle\langle d_{\sigma} | d^{\dagger}_{\sigma} \rangle\rangle^{r}_{\varepsilon} = 1 + U \langle\langle d_{\sigma} n_{d\bar{\sigma}} | d^{\dagger}_{\sigma} \rangle\rangle^{r}_{\varepsilon}. \tag{3.40}
$$

The EOM for the term $\langle\langle d_{\sigma} n_{d\bar{\sigma}} | d^{\dagger}_{\sigma} \rangle\rangle^{r}_{\varepsilon}$ reads

$$
\varepsilon^{+} \langle\langle d_{\sigma} n_{d\bar{\sigma}} | d^{\dagger}_{\sigma} \rangle\rangle^{r}_{\varepsilon} + \langle\langle [\mathcal{H}, d_{\sigma} n_{d\bar{\sigma}}] | d^{\dagger}_{\sigma} \rangle\rangle^{r}_{\varepsilon} = \langle \{ d_{\sigma} n_{d\bar{\sigma}}, d^{\dagger}_{\sigma} \} \rangle, \tag{3.41}
$$

where

$$
\{ d_{\sigma} n_{d\bar{\sigma}}, d^{\dagger}_{\sigma} \} = \underbrace{\{d^{\dagger}_{\sigma}, d_{\sigma}\}}_{=1} n_{d\bar{\sigma}} - d_{\sigma} \big(\underbrace{\{d^{\dagger}_{\sigma}, d^{\dagger}_{\bar{\sigma}}\}}_{=0} d_{\bar{\sigma}} - d^{\dagger}_{\bar{\sigma}} \underbrace{\{d^{\dagger}_{\sigma}, d_{\bar{\sigma}}\}}_{=0} \big) = n_{d\bar{\sigma}}, \tag{3.42}
$$

$$
[\mathcal{H}, d_{\sigma} n_{d\bar{\sigma}}] = [\mathcal{H}_{\text{Leads}}, d_{\sigma} n_{d\bar{\sigma}}] + [\mathcal{H}_{\text{QD}}, d_{\sigma} n_{d\bar{\sigma}}] + [\mathcal{H}_{\text{T}}, d_{\sigma} n_{d\bar{\sigma}}], \tag{3.43}
$$

$$
\begin{aligned}
[\mathcal{H}_{\text{Leads}}, d_\sigma n_{d\bar{\sigma}}] &= \sum_{\alpha,s,\eta} \int_{-k_c}^{+k_c} dk\, \varepsilon_k \left[c^\dagger_{\alpha s k \eta} c_{\alpha s k \eta}, d_\sigma n_{d\bar{\sigma}}\right] \\
&= \sum_{\alpha,s,\eta} \int_{-k_c}^{+k_c} dk\, \varepsilon_k \big(\left[c^\dagger_{\alpha s k \eta} c_{\alpha s k \eta}, d_\sigma\right] n_{d\bar{\sigma}} + d_\sigma \left[c^\dagger_{\alpha s k \eta} c_{\alpha s k \eta}, n_{d\bar{\sigma}}\right]\big) \\
&= -\sum_{\alpha,s,\eta} \int_{-k_c}^{+k_c} dk\, \varepsilon_k \Big[\big(\underbrace{\{d_\sigma, c^\dagger_{\alpha s k \eta}\}}_{=0} c_{\alpha s k \eta} - c^\dagger_{\alpha s k \eta} \underbrace{\{d_\sigma, c_{\alpha s k \eta}\}}_{=0}\big) n_{d\bar{\sigma}} \\
&\quad - d_\sigma \big(\underbrace{\{c^\dagger_{\alpha s k \eta}, d^\dagger_{\bar{\sigma}}\}}_{=0} d_{\bar{\sigma}} - d^\dagger_{\bar{\sigma}} \underbrace{\{c^\dagger_{\alpha s k \eta}, d_{\bar{\sigma}}\}}_{=0}\big) c_{\alpha s k \eta} \\
&\quad - d_\sigma c^\dagger_{\alpha s k \eta} \big(\underbrace{\{c_{\alpha s k \eta}, d^\dagger_{\bar{\sigma}}\}}_{=0} d_{\bar{\sigma}} - d^\dagger_{\bar{\sigma}} \underbrace{\{c_{\alpha s k \eta}, d_{\bar{\sigma}}\}}_{=0}\big)\Big] = 0,
\end{aligned} \tag{3.44}
$$

$$
\begin{aligned}
[\mathcal{H}_{\text{QD}}, d_\sigma n_{d\bar{\sigma}}] &= \sum_\eta \varepsilon_{d\eta} \left[d^\dagger_\eta d_\eta, d_\sigma n_{d\bar{\sigma}}\right] + U \left[n_{d\uparrow} n_{d\downarrow}, d_\sigma n_{d\bar{\sigma}}\right] \\
&= -\sum_\eta \varepsilon_{d\eta} \big(\left[d_\sigma, d^\dagger_\eta d_\eta\right] n_{d\bar{\sigma}} + d_\sigma \left[d^\dagger_{\bar{\sigma}}, d^\dagger_\eta d_\eta\right] d_{\bar{\sigma}} + d_\sigma d^\dagger_{\bar{\sigma}} \left[d_{\bar{\sigma}}, d^\dagger_\eta d_\eta\right]\big) \\
&\quad - U \big(\left[d_\sigma, n_{d\uparrow} n_{d\downarrow}\right] n_{d\bar{\sigma}} + d_\sigma \left[n_{d\bar{\sigma}}, n_{d\uparrow} n_{d\downarrow}\right]\big) \\
&= -\sum_\eta \varepsilon_{d\eta} \Big[\big(\underbrace{\{d_\sigma, d^\dagger_\eta\}}_{=\delta_{\sigma\eta}} d_\eta - d^\dagger_\eta \underbrace{\{d_\sigma, d_\eta\}}_{=0}\big) n_{d\bar{\sigma}} \\
&\quad + d_\sigma \big(\underbrace{\{d^\dagger_{\bar{\sigma}}, d^\dagger_\eta\}}_{=0} d_\eta - d^\dagger_\eta \underbrace{\{d^\dagger_{\bar{\sigma}}, d_\eta\}}_{=\delta_{\eta\bar{\sigma}}}\big) d_{\bar{\sigma}} + d_\sigma d^\dagger_{\bar{\sigma}} \big(\underbrace{\{d_{\bar{\sigma}}, d^\dagger_\eta\}}_{=\delta_{\bar{\sigma}\eta}} d_\eta - d^\dagger_\eta \underbrace{\{d_{\bar{\sigma}}, d_\eta\}}_{=0}\big)\Big] \\
&\quad - U \big([d_\sigma, d^\dagger_\uparrow d_\uparrow] n_{d\downarrow} n_{d\bar{\sigma}} + n_{d\uparrow} [d_\sigma, d^\dagger_\downarrow d_\downarrow] n_{d\bar{\sigma}} + d_\sigma [n_{d\bar{\sigma}}, d^\dagger_\uparrow] d_\uparrow n_{d\downarrow} \\
&\quad + d_\sigma d^\dagger_\uparrow [n_{d\bar{\sigma}}, d_\uparrow] n_{d\downarrow} + d_\sigma n_{d\uparrow} [n_{d\bar{\sigma}}, d^\dagger_\downarrow] d_\downarrow + d_\sigma n_{d\uparrow} d^\dagger_\downarrow [n_{d\bar{\sigma}}, d_\downarrow]\big) = -\varepsilon_{d\sigma}\, d_\sigma n_{d\bar{\sigma}} \\
&\quad - U \Big[\big(\underbrace{\{d_\sigma, d^\dagger_\uparrow\}}_{=\delta_{\sigma\uparrow}} d_\uparrow - d^\dagger_\uparrow \underbrace{\{d_\sigma, d_\uparrow\}}_{=0}\big) n_{d\downarrow} n_{d\bar{\sigma}} + n_{d\uparrow} \big(\underbrace{\{d_\sigma, d^\dagger_\downarrow\}}_{=\delta_{\sigma\downarrow}} d_\downarrow - d^\dagger_\downarrow \underbrace{\{d_\sigma, d_\downarrow\}}_{=0}\big) n_{d\bar{\sigma}} \\
&\quad - d_\sigma \big(\underbrace{\{d^\dagger_\uparrow, d^\dagger_{\bar{\sigma}}\}}_{=0} d_{\bar{\sigma}} - d^\dagger_{\bar{\sigma}} \underbrace{\{d^\dagger_\uparrow, d_{\bar{\sigma}}\}}_{=\delta_{\bar{\sigma}\uparrow}}\big) d_\uparrow n_{d\downarrow} - d_\sigma d^\dagger_\uparrow \big(\underbrace{\{d_\uparrow, d^\dagger_{\bar{\sigma}}\}}_{=\delta_{\uparrow\bar{\sigma}}} d_{\bar{\sigma}} - d^\dagger_{\bar{\sigma}} \underbrace{\{d_\uparrow, d_{\bar{\sigma}}\}}_{=0}\big) n_{d\downarrow} \\
&\quad - d_\sigma n_{d\uparrow} \big(\underbrace{\{d^\dagger_\downarrow, d^\dagger_{\bar{\sigma}}\}}_{=0} d_{\bar{\sigma}} - d^\dagger_{\bar{\sigma}} \underbrace{\{d^\dagger_\downarrow, d_{\bar{\sigma}}\}}_{=\delta_{\bar{\sigma}\downarrow}}\big) d_\downarrow - d_\sigma n_{d\uparrow} d^\dagger_\downarrow \big(\underbrace{\{d_\downarrow, d^\dagger_{\bar{\sigma}}\}}_{=\delta_{\downarrow\bar{\sigma}}} d_{\bar{\sigma}} - d^\dagger_{\bar{\sigma}} \underbrace{\{d_\downarrow, d_{\bar{\sigma}}\}}_{=0}\big)\Big] \\
&= -\varepsilon_{d\sigma}\, d_\sigma n_{d\bar{\sigma}} - U\, d_\sigma n_{d\bar{\sigma}},
\end{aligned} \tag{3.45}
$$

and

$$
\begin{aligned}
[\mathcal{H}_{\text{T}}, d_\sigma n_{d\bar{\sigma}}] &= \sum_{\alpha,s,\eta} \int_{-k_c}^{+k_c} dk \big(\mathcal{V}_{\alpha s\eta}(k) [c^\dagger_{\alpha s k \eta} d_\eta, d_\sigma n_{d\bar{\sigma}}] + [\mathcal{V}_{\alpha s\eta}(k)]^* [d^\dagger_\eta c_{\alpha s k \eta}, d_\sigma n_{d\bar{\sigma}}]\big) \\
&= -\sum_{\alpha,s,\eta} \int_{-k_c}^{+k_c} dk\, \mathcal{V}_{\alpha s\eta}(k) \Big(\left[d_\sigma, c^\dagger_{\alpha s k \eta} d_\eta\right] n_{d\bar{\sigma}} + d_\sigma \left[n_{d\bar{\sigma}}, c^\dagger_{\alpha s k \eta} d_\eta\right]\Big)
\end{aligned}
$$

$$
\begin{aligned}
&-\sum_{\alpha,s,\eta}\int_{-k_c}^{+k_c} dk[\mathcal{V}_{\alpha s\eta}(k)]^*\Big([d_\sigma, d_\eta^\dagger c_{\alpha sk\eta}]n_{d\bar{\sigma}} + d_\sigma[n_{d\bar{\sigma}}, d_\eta^\dagger c_{\alpha sk\eta}]\Big)\\
&= -\sum_{\alpha,s,\eta}\int_{-k_c}^{+k_c} dk\,\mathcal{V}_{\alpha s\eta}(k)\Big[\big(\underbrace{\{d_\sigma, c_{\alpha sk\eta}^\dagger\}}_{=0}d_\eta - c_{\alpha sk\eta}^\dagger\underbrace{\{d_\sigma, d_\eta\}}_{=0}\big)n_{d\bar{\sigma}}\\
&- d_\sigma\big(\underbrace{\{c_{\alpha sk\eta}^\dagger, d_{\bar{\sigma}}^\dagger\}}_{=0}d_{\bar{\sigma}} - d_{\bar{\sigma}}^\dagger\underbrace{\{c_{\alpha sk\eta}^\dagger, d_{\bar{\sigma}}\}}_{=0}\big)d_\eta - d_\sigma c_{\alpha sk\eta}^\dagger\big(\underbrace{\{d_\eta, d_{\bar{\sigma}}^\dagger\}}_{=\delta_{\eta\bar{\sigma}}}d_{\bar{\sigma}} - d_{\bar{\sigma}}^\dagger\underbrace{\{d_\eta, d_{\bar{\sigma}}\}}_{=0}\big)\Big]\\
&-\sum_{\alpha,s,\eta}\int_{-k_c}^{+k_c} dk[\mathcal{V}_{\alpha s\eta}(k)]^*\Big[\big(\underbrace{\{d_\sigma, d_\eta^\dagger\}}_{=\delta_{\sigma\eta}}c_{\alpha sk\eta} - d_\eta^\dagger\underbrace{\{d_\sigma, c_{\alpha sk\eta}\}}_{=0}\big)n_{d\bar{\sigma}}\\
&- d_\sigma\big(\underbrace{\{d_\eta^\dagger, d_{\bar{\sigma}}^\dagger\}}_{=0}d_{\bar{\sigma}} - d_{\bar{\sigma}}^\dagger\underbrace{\{d_\eta^\dagger, d_{\bar{\sigma}}\}}_{=\delta_{\bar{\sigma}\eta}}\big)c_{\alpha sk\eta} - d_\sigma d_\eta^\dagger\big(\underbrace{\{c_{\alpha sk\eta}, d_{\bar{\sigma}}^\dagger\}}_{=0}d_{\bar{\sigma}} - d_{\bar{\sigma}}^\dagger\underbrace{\{c_{\alpha sk\eta}, d_{\bar{\sigma}}\}}_{=0}\big)\Big]\\
&= \sum_{\alpha,s}\int_{-k_c}^{+k_c} dk\big(\mathcal{V}_{\alpha s\bar{\sigma}}(k)\,d_\sigma c_{\alpha sk\bar{\sigma}}^\dagger d_{\bar{\sigma}} - [\mathcal{V}_{\alpha s\sigma}(k)]^* c_{\alpha sk\sigma}n_{d\bar{\sigma}} - [\mathcal{V}_{\alpha s\bar{\sigma}}(k)]^* d_\sigma d_{\bar{\sigma}}^\dagger c_{\alpha sk\bar{\sigma}}\big).
\end{aligned}
\tag{3.46}
$$

Substituting relations from (3.44)–(3.46) into Eq. (3.43), and using relation (3.42), Eq. (3.41) reads

$$
\begin{aligned}
(\varepsilon^+ - \varepsilon_{d\sigma} - U)\langle\langle d_\sigma n_{d\bar{\sigma}}|d_\sigma^\dagger\rangle\rangle_\varepsilon^r &= \langle n_{d\bar{\sigma}}\rangle + \sum_{\alpha,s}\int_{-k_c}^{+k_c} dk\Big[[\mathcal{V}_{\alpha s\sigma}(k)]^*\langle\langle c_{\alpha sk\sigma}n_{d\bar{\sigma}}|d_\sigma^\dagger\rangle\rangle_\varepsilon^r\\
&+ [\mathcal{V}_{\alpha s\bar{\sigma}}(k)]^*\langle\langle d_{\bar{\sigma}}^\dagger c_{\alpha sk\bar{\sigma}}d_\sigma|d_\sigma^\dagger\rangle\rangle_\varepsilon^r - \mathcal{V}_{\alpha s\bar{\sigma}}(k)\langle\langle c_{\alpha sk\bar{\sigma}}^\dagger d_{\bar{\sigma}}d_\sigma|d_\sigma^\dagger\rangle\rangle_\varepsilon^r\Big].
\end{aligned}
\tag{3.47}
$$

To determine the dot Green's function, we need to calculate the new higher-order correlation functions that appear on the right-hand side of Eq. (3.47). Following the steps presented above, the equations of motion for these terms are expressed as follows:

$$
\begin{aligned}
\Omega_k^{(0)}(\varepsilon)\langle\langle c_{\alpha sk\sigma}n_{d\bar{\sigma}}|d_\sigma^\dagger\rangle\rangle_\varepsilon^r &= \mathcal{V}_{\alpha s\sigma}(k)\langle\langle d_\sigma n_{d\bar{\sigma}}|d_\sigma^\dagger\rangle\rangle_\varepsilon^r\\
-\sum_{\beta,r}\int_{-k_c}^{+k_c} dq\Big[\mathcal{V}_{\beta r\bar{\sigma}}(q)\langle\langle c_{\alpha sk\sigma}c_{\beta rq\bar{\sigma}}^\dagger d_{\bar{\sigma}}|d_\sigma^\dagger\rangle\rangle_\varepsilon^r &+ [\mathcal{V}_{\beta r\bar{\sigma}}(q)]^*\langle\langle c_{\alpha sk\sigma}c_{\beta rq\bar{\sigma}}d_{\bar{\sigma}}^\dagger|d_\sigma^\dagger\rangle\rangle_\varepsilon^r\Big],
\end{aligned}
\tag{3.48}
$$

$$
\begin{aligned}
\Omega_{k\sigma}^{(1)}(\varepsilon)\langle\langle d_{\bar{\sigma}}^\dagger c_{\alpha sk\bar{\sigma}}d_\sigma|d_\sigma^\dagger\rangle\rangle_\varepsilon^r &= \langle d_{\bar{\sigma}}^\dagger c_{\alpha sk\bar{\sigma}}\rangle + \mathcal{V}_{\alpha s\bar{\sigma}}(k)\langle\langle d_\sigma n_{d\bar{\sigma}}|d_\sigma^\dagger\rangle\rangle_\varepsilon^r\\
+\sum_{\beta,r}\int_{-k_c}^{+k_c} dq\Big[[\mathcal{V}_{\beta r\sigma}(q)]^*\langle\langle d_{\bar{\sigma}}^\dagger c_{\alpha sk\bar{\sigma}}c_{\beta rq\sigma}|d_\sigma^\dagger\rangle\rangle_\varepsilon^r &- \mathcal{V}_{\beta r\bar{\sigma}}(q)\langle\langle c_{\beta rq\bar{\sigma}}^\dagger c_{\alpha sk\bar{\sigma}}d_\sigma|d_\sigma^\dagger\rangle\rangle_\varepsilon^r\Big],
\end{aligned}
\tag{3.49}
$$

$$
\begin{aligned}
\Omega_{k\sigma}^{(2)}(\varepsilon)\langle\langle c_{\alpha sk\bar{\sigma}}^\dagger d_{\bar{\sigma}}d_\sigma|d_\sigma^\dagger\rangle\rangle_\varepsilon^r &= \langle c_{\alpha sk\bar{\sigma}}^\dagger d_{\bar{\sigma}}\rangle - [\mathcal{V}_{\alpha s\bar{\sigma}}(k)]^*\langle\langle d_\sigma n_{d\bar{\sigma}}|d_\sigma^\dagger\rangle\rangle_\varepsilon^r\\
+\sum_{\beta,r}\int_{-k_c}^{+k_c} dq\Big[[\mathcal{V}_{\beta r\bar{\sigma}}(q)]^*\langle\langle c_{\alpha sk\bar{\sigma}}^\dagger c_{\beta rq\bar{\sigma}}d_\sigma|d_\sigma^\dagger\rangle\rangle_\varepsilon^r &- [\mathcal{V}_{\beta r\sigma}(q)]^*\langle\langle c_{\alpha sk\bar{\sigma}}^\dagger c_{\beta rq\sigma}d_{\bar{\sigma}}|d_\sigma^\dagger\rangle\rangle_\varepsilon^r\Big],
\end{aligned}
\tag{3.50}
$$

where the following notations are used: $\Omega_k^{(0)}(\varepsilon) \equiv \varepsilon^+ - \varepsilon_k$, $\Omega_{k\sigma}^{(1)}(\varepsilon) \equiv \varepsilon^+ - \varepsilon_k + (\varepsilon_{d\bar{\sigma}} - \varepsilon_{d\sigma})$ and $\Omega_{k\sigma}^{(2)}(\varepsilon) \equiv \varepsilon^+ + \varepsilon_k - (\varepsilon_{d\bar{\sigma}} + \varepsilon_{d\sigma}) - U$. To obtain an analytical formula for the dot Green's function, we have to truncate the higher-order correlation functions that appear in Eqs. (3.48)–(3.50) by using an approximation method. In order to do this, we apply the broadly used Lacroix decoupling scheme [41] that leads to close the infinite number of higher-order correlation functions. By performing the approximations and substituting the resulting equations into Eq. (3.47) for the dot Green's function we obtain

$$G_{d\sigma}^r(\varepsilon) = \frac{\Pi_\sigma^{(1)}(\varepsilon) + U\big[\langle n_{d\bar{\sigma}}\rangle + \Pi_\sigma^{(2)}(\varepsilon)\big]}{\Pi_\sigma^{(1)}(\varepsilon)\big[\varepsilon - \varepsilon_{d\sigma} - \Sigma_0^r(\varepsilon)\big] - U\Pi_\sigma^{(3)}(\varepsilon)}, \tag{3.51}$$

where we introduced the notations:

$$\Pi_\sigma^{(1)}(\varepsilon) = \varepsilon - \varepsilon_{d\sigma} - U - \Sigma_0^r(\varepsilon) - \sum_{\alpha,s}\int_{-k_c}^{+k_c} dk\, |\mathcal{V}_{\alpha s\bar{\sigma}}(k)|^2\left[\frac{1}{\Omega_{k\sigma}^{(1)}} + \frac{1}{\Omega_{k\sigma}^{(2)}}\right], \tag{3.52}$$

$$\Pi_\sigma^{(2)}(\varepsilon) = \sum_{\alpha,s}\int_{-k_c}^{+k_c} dk\left[[\mathcal{V}_{\alpha s\bar{\sigma}}(k)]^*\frac{\langle d_{\bar{\sigma}}^\dagger c_{\alpha s k\bar{\sigma}}\rangle}{\Omega_{k\sigma}^{(1)}} - \mathcal{V}_{\alpha s\bar{\sigma}}(k)\frac{\langle c_{\alpha s k\bar{\sigma}}^\dagger d_{\bar{\sigma}}\rangle}{\Omega_{k\sigma}^{(2)}}\right], \tag{3.53}$$

$$\Pi_\sigma^{(3)}(\varepsilon) = \Sigma_0^r(\varepsilon)\Pi_\sigma^{(2)}(\varepsilon)$$
$$-\sum_{\substack{\alpha,s\\ \beta,r}}\iint_{-k_c}^{+k_c} dkdq\left[[\mathcal{V}_{\alpha s\bar{\sigma}}(k)]^*\mathcal{V}_{\beta r\bar{\sigma}}(q)\frac{\langle c_{\beta r q\bar{\sigma}}^\dagger c_{\alpha s k\bar{\sigma}}\rangle}{\Omega_{k\sigma}^{(1)}} + \mathcal{V}_{\alpha s\bar{\sigma}}(k)[\mathcal{V}_{\beta r\bar{\sigma}}(q)]^*\frac{\langle c_{\alpha s k\bar{\sigma}}^\dagger c_{\beta r q\bar{\sigma}}\rangle}{\Omega_{k\sigma}^{(2)}}\right]. \tag{3.54}$$

The average values of mixing operators in the above relations are treated non-self-consistently, following the Meir approximation [10], i.e., $\langle d_{\bar{\sigma}}^\dagger c_{\alpha s k\bar{\sigma}}\rangle = \langle c_{\alpha s k\bar{\sigma}}^\dagger d_{\bar{\sigma}}\rangle \approx 0$ and $\langle c_{\beta r q\bar{\sigma}}^\dagger c_{\alpha s k\bar{\sigma}}\rangle = \langle c_{\alpha s k\bar{\sigma}}^\dagger c_{\beta r q\bar{\sigma}}\rangle \approx f_\alpha(\varepsilon_k)\delta_{\alpha\beta}\delta_{sr}\delta(k-q)$, leading to $\Pi_\sigma^{(2)}(\varepsilon) \approx 0$. Applying these approximations with $\mathcal{V}_{\alpha s\bar{\sigma}}(k) = \mathcal{V}(k)$, the form of the retarded Green's function given by Eq. (3.51) reduces to the form given by Eq. (3.9) with relations (3.10)–(3.13).

References

1. K.S. Novoselov, A.K. Geim, S.V. Morozov, D. Jiang, Y. Zhang, S.V. Dubonos, I.V. Grigorieva, A.A. Firsov, Electric Field Effect in Atomically Thin Carbon Films. Science **306**, 666–669 (2004)
2. K.S. Novoselov, A.K. Geim, S.V. Morozov, D. Jiang, M.I. Katsnelson, I.V. Grigorieva, S.V. Dubonos, A.A. Firsov, Two-dimensional gas of massless Dirac fermions in graphene. Nature **438**, 197–200 (2005)
3. Z. Zhang, J. Zhang, N. Chen, L. Qu, Graphene quantum dots: an emerging material for energy-related applications and beyond. Energy Environ. Sci. **5**, 8869–8890 (2012)
4. H. Sun, L. Wu, W. Wei, X. Qu, Recent advances in graphene quantum dots for sensing. Mater. Today **16**(11), 433–442 (2013)

5. S.-H. Choi, Unique properties of graphene quantum dots and their applications in photonic/electronic devices. J. Phys. D Appl. Phys. **50**(10), 103002 (2017)
6. T.-H. Han, H. Kim, S.-J. Kwon, T.-W. Lee, Graphene-based flexible electronic devices. Mater. Sci. Eng. R. Rep. **118**, 1–43 (2017)
7. A. Rozhkov, G. Giavaras, Y.P. Bliokh, V. Freilikher, F. Nori, Electronic properties of mesoscopic graphene structures: Charge confinement and control of spin and charge transport. Phys. Rep. **503**, 77–114 (2011)
8. A.H. Castro Neto, F. Guinea, N.M.R. Peres, K.S. Novoselov, A.K. Geim, The electronic properties of graphene. Rev. Mod. Phys. **81**, 109–162 (2009)
9. V.N. Kotov, B. Uchoa, V.M. Pereira, F. Guinea, A.H. Castro Neto, Electron-electron interactions in graphene: Current status and perspectives. Rev. Mod. Phys. **84**, 1067–1125 (2012)
10. Y. Meir, N.S. Wingreen, P.A. Lee, Transport through a strongly interacting electron system: Theory of periodic conductance oscillations. Phys. Rev. Lett. **66**, 3048–3051 (1991)
11. Y. Meir, N.S. Wingreen, Landauer formula for the current through an interacting electron region. Phys. Rev. Lett. **68**, 2512–2515 (1992)
12. S. Hershfield, J.H. Davies, J.W. Wilkins, Resonant tunneling through an Anderson impurity. I. Current in the symmetric model. Phys. Rev. B **46**, 7046–7060 (1992)
13. Y. Meir, N.S. Wingreen, P.A. Lee, Low-temperature transport through a quantum dot: The Anderson model out of equilibrium. Phys. Rev. Lett. **70**, 2601–2604 (1993)
14. A.L. Yeyati, A. Martín-Rodero, F. Flores, Electron correlation resonances in the transport through a single quantum level. Phys. Rev. Lett. **71**, 2991–2994 (1993)
15. N.S. Wingreen, Y. Meir, Anderson model out of equilibrium: Noncrossing-approximation approach to transport through a quantum dot. Phys. Rev. B **49**, 11040–11052 (1994)
16. A.-P. Jauho, N.S. Wingreen, Y. Meir, Time-dependent transport in interacting and noninteracting resonant-tunneling systems. Phys. Rev. B **50**, 5528–5544 (1994)
17. T. Aono, M. Eto, K. Kawamura, Conductance through Quantum Dot Dimer Below the Kondo Temperature. J. Phys. Soc. Jpn. **67**(6), 1860–1863 (1998)
18. B. Dong, X.L. Lei, Kondo-type transport through a quantum dot under magnetic fields. Phys. Rev. B **63**, 235306 (2001)
19. M. Krawiec, K.I. Wysokiński, Nonequilibrium Kondo effect in asymmetrically coupled quantum dots. Phys. Rev. B **66**, 165408 (2002)
20. R. Świrkowicz, J. Barnaś, M. WilczyÇ¹ski, Nonequilibrium Kondo effect in quantum dots. Phys. Rev. B **68**, 195318 (Nov2003)
21. A. C. Hewson, J. Bauer, A. Oguri, Non-equilibrium differential conductance through a quantum dot in a magnetic field. J. Phys.: Condens. Matter. **17**(35), 5413 (2005)
22. C.-H. Chung, G. Zarand, P. Wölfle, Two-stage Kondo effect in side-coupled quantum dots: Renormalized perturbative scaling theory and numerical renormalization group analysis. Phys. Rev. B **77**, 035120 (2008)
23. N.A. Zimbovskaya, Electron transport through a quantum dot in the Coulomb blockade regime: Nonequilibrium Green's function based model. Phys. Rev. B **78**, 035331 (2008)
24. S. Sasaki, H. Tamura, T. Akazaki, T. Fujisawa, Fano-Kondo interplay in a side-coupled double quantum dot. Phys. Rev. Lett. **103**, 266806 (2009)
25. R. Van Roermund, S.-Y. Shiau, M. Lavagna, Anderson model out of equilibrium: Decoherence effects in transport through a quantum dot. Phys. Rev. B **81**, 165115 (2010)
26. N.A. Zimbovskaya, The effect of Coulomb interactions on thermoelectric properties of quantum dots. J. Chem. Phys. **140**(10), 104706 (2014)
27. N.A. Zimbovskaya, The effect of Coulomb interactions on nonlinear thermovoltage and thermocurrent in quantum dots. J. Chem. Phys. **142**(24), 244310 (2015)
28. M. Crisan, I. Grosu, I. Tifrea, An equation of motion analysis of the two stage Kondo effect in T-shaped double-quantum-dot systems. Physica E **66**, 245–251 (2015)
29. S. Sahoo, A. Crpieux, M. Lavagna, Out-of-equilibrium Kondo effect in a quantum dot: Interplay of magnetic field and spin accumulation. EPL **116**(5), 57005 (2016)

30. M.A. Sierra, R. López, D. Sánchez, Fate of the spin-$\frac{1}{2}$ Kondo effect in the presence of temperature gradients. Phys. Rev. B **96**, 085416 (2017)
31. J. Martinek, Y. Utsumi, H. Imamura, J. Barnaś, S. Maekawa, J. König, G. Schön, Kondo effect in quantum dots coupled to ferromagnetic leads. Phys. Rev. Lett. **91**, 127203 (2003)
32. M.-S. Choi, D. Sánchez, R. López, Kondo effect in a quantum dot coupled to ferromagnetic leads: A numerical renormalization group analysis. Phys. Rev. Lett. **92**, 056601 (2004)
33. P. Trocha, J. Barnaś, Quantum interference and Coulomb correlation effects in spin-polarized transport through two coupled quantum dots. Phys. Rev. B **76**, 165432 (2007)
34. R. Žitko, J.S. Lim, R. López, J. Martinek, P. Simon, Tunable Kondo effect in a double quantum dot coupled to ferromagnetic contacts. Phys. Rev. Lett. **108**, 166605 (2012)
35. T. Aono, Electrical and thermoelectrical transport properties of Dirac fermions through a quantum dot. J. Phys. Soc. Jpn. **82**(8), 083703 (2013)
36. T. Aono, Conductance and thermopower of Dirac fermions under the Kondo effect. JPS Conf. Proc. **3**, 012022 (2014)
37. T. Aono, T. Komine, Giant thermoelectric figure of merit in a noninteracting quantum dot system with massless Dirac fermions. Phys. Rev. B **94**, 165311 (2016)
38. S. Rodriguez, I. Grosu, M. Crisan, I. Tifrea, Thermoelectric transport properties in graphene connected molecular junctions. Physica E **96**, 1–5 (2018)
39. J.R. Isern-Lozano, J.S. Lim, I. Grosu, R. López, M. Crisan, D. Sánchez, Thermoelectric transport through interacting quantum dots in graphene. Eur. Phys. J. Spec. Top. **227**, 1969–1979 (2019)
40. P.W. Anderson, Localized magnetic states in metals. Phys. Rev. **124**, 41–53 (1961)
41. C. Lacroix, Density of states for the Anderson model. J. Phys. F: Met. Phys. **11**, 2389–2397 (1981)
42. S.M. Cronenwett, T.H. Oosterkamp, L.P. Kouwenhoven, A tunable Kondo effect in quantum dots. Science **281**(5376), 540–544 (1998)
43. W.G. van der Wiel, S.D. Franceschi, T. Fujisawa, J.M. Elzerman, S. Tarucha, L.P. Kouwenhoven, The Kondo effect in the unitary limit. Science **289**(5487), 2105–2108 (2000)
44. R.C. Monreal, F. Flores, Kondo resonance decoherence caused by an external potential. Phys. Rev. B **72** (2005)
45. V. Kashcheyevs, A. Aharony, O. Entin-Wohlman, Applicability of the equations-of-motion technique for quantum dots. Phys. Rev. B **73**, 125338 (2006)
46. J.S. Lim, R. López, L. Limot, P. Simon, Nonequilibrium spin-current detection with a single Kondo impurity. Phys. Rev. B **88**, 165403 (2013)
47. D. Gerace, E. Pavarini, L.C. Andreani, Low-temperature transport through a quantum dot: Finite-u results and scaling behavior. Phys. Rev. B **65**, 155331 (2002)
48. M. Krawiec, K.I. Wysokiński, Thermoelectric phenomena in a quantum dot asymmetrically coupled to external leads. Phys. Rev. B **75**, 155330 (2007)
49. A.A. Aligia, P. Roura-Bas, S. Florens, Impact of capacitance and tunneling asymmetries on Coulomb blockade edges and Kondo peaks in nonequilibrium transport through molecular quantum dots. Phys. Rev. B **92**, 035404 (2015)
50. D. Pérez Daroca, P. Roura-Bas, A.A. Aligia, Enhancing the nonlinear thermoelectric response of a correlated quantum dot in the Kondo regime by asymmetrical coupling to the leads. Phys. Rev. B **97**, 165433 (2018)
51. D.P. Daroca, P. Roura-Bas, A.A. Aligia, Relation between width of zero-bias anomaly and Kondo temperature in transport measurements through correlated quantum dots: Effect of asymmetric coupling to the leads. Phys. Rev. B **98**, 245406 (2018)
52. W. Izumida, O. Sakai, Y. Shimizu, Kondo effect in single quantum dot systems – study with numerical renormalization group method –. J. Phys. Soc. Jpn. **67**(7), 2444–2454 (1998)
53. D.M. Kennes, S.G. Jakobs, C. Karrasch, V. Meden, Renormalization group approach to time-dependent transport through correlated quantum dots. Phys. Rev. B **85**, 085113 (2012)
54. M.A. Manya, G.B. Martins, M.S. Figueira, Spin-orbit coupling effects on thermoelectric transport properties in quantum dots. Phys. Rev. B **105**, 165421 (2022)

55. A. Manaparambil, I. Weymann, Nonequilibrium Seebeck effect and thermoelectric efficiency of Kondo-correlated molecular junctions. Phys. Rev. B **107**, 085404 (2023)
56. J.E. Han, R.J. Heary, Imaginary-time formulation of steady-state nonequilibrium: Application to strongly correlated transport. Phys. Rev. Lett. **99**, 236808 (2007)
57. J.E. Han, Nonequilibrium electron transport in strongly correlated molecular junctions. Phys. Rev. B **81**, 113106 (2010)
58. A. Georges, Y. Meir, Electronic correlations in transport through coupled quantum dots. Phys. Rev. Lett. **82**, 3508–3511 (1999)
59. R. Aguado, D.C. Langreth, Out-of-equilibrium Kondo effect in double quantum dots. Phys. Rev. Lett. **85**, 1946–1949 (2000)
60. A.J. Keller, S. Amasha, I. Weymann, C.P. Moca, I.G. Rau, J.A. Katine, H. Shtrikman, G. Zaránd, D. Goldhaber-Gordon, Emergent SU(4) Kondo physics in a spin–charge-entangled double quantum dot. Nature Phys. **10**, 145–150 (2014)
61. T.A. Costi, Magnetic field dependence of the thermopower of Kondo-correlated quantum dots. Phys. Rev. B **100** (2019)
62. D.C. Ralph, R.A. Buhrman, Kondo-assisted and resonant tunneling via a single charge trap: A realization of the Anderson model out of equilibrium. Phys. Rev. Lett. **72**, 3401–3404 (1994)
63. D. Goldhaber-Gordon, H. Shtrikman, D. Mahalu, D. Abusch-Magder, U. Meirav, M.A. Kastner, Kondo effect in a single-electron transistor. Nature **391**, 156–159 (1998)
64. D. Goldhaber-Gordon, J. Göres, M.A. Kastner, H. Shtrikman, D. Mahalu, U. Meirav, From the Kondo regime to the mixed-valence regime in a single-electron transistor. Phys. Rev. Lett. **81**, 5225–5228 (1998)
65. J. Schmid, J. Weis, K. Eberl, K.V. Klitzing, A quantum dot in the limit of strong coupling to reservoirs. Phys. B **256-258**, 182–185 (1998)
66. F. Simmel, R.H. Blick, J.P. Kotthaus, W. Wegscheider, M. Bichler, Anomalous Kondo effect in a quantum dot at nonzero bias. Phys. Rev. Lett. **83**, 804–807 (1999)
67. J. Schmid, J. Weis, K. Eberl, K.V. Klitzing, Absence of odd-even parity behavior for Kondo resonances in quantum dots. Phys. Rev. Lett. **84**, 5824–5827 (2000)
68. J. Schmid, J. Weis, K. Eberl, K. von Klitzing, Split Kondo resonances in quantum dots at finite magnetic fields. Phys. E **9**(1), 54–59 (2001). Proceedings of the Eleventh International Winterschool on New Developments in Solid State Physics, "New Frontiers in Low-Dimensional Physics"
69. S. De Franceschi, R. Hanson, W.G. van der Wiel, J.M. Elzerman, J.J. Wijpkema, T. Fujisawa, S. Tarucha, L.P. Kouwenhoven, Out-of-equilibrium Kondo effect in a mesoscopic device. Phys. Rev. Lett. **89**, 156801 (2002)
70. A. Kogan, S. Amasha, D. Goldhaber-Gordon, G. Granger, M.A. Kastner, H. Shtrikman, Measurements of Kondo and spin splitting in single-electron transistors. Phys. Rev. Lett. **93**, 166602 (2004)
71. L.J. Klein, D.E. Savage, M.A. Eriksson, Coulomb blockade and Kondo effect in a few-electron silicon/silicon-germanium quantum dot. Appl. Phys. Lett. **90**, 033103, 01 (2007)
72. M. Grobis, I.G. Rau, R.M. Potok, H. Shtrikman, D. Goldhaber-Gordon, Universal scaling in nonequilibrium transport through a single channel Kondo dot. Phys. Rev. Lett. **100**, 246601 (2008)
73. A.V. Kretinin, H. Shtrikman, D. Goldhaber-Gordon, M. Hanl, A. Weichselbaum, J. von Delft, T. Costi, D. Mahalu, Spin-$\frac{1}{2}$ Kondo effect in an InAs nanowire quantum dot: Unitary limit, conductance scaling, and Zeeman splitting. Phys. Rev. B **84**, 245316 (2011)
74. O. Klochan, A.P. Micolich, A.R. Hamilton, D. Reuter, A.D. Wieck, F. Reininghaus, M. Pletyukhov, H. Schoeller, Scaling of the Kondo zero-bias peak in a hole quantum dot at finite temperatures. Phys. Rev. B **87**, 201104 (2013)
75. A. Svilans, M. Josefsson, A.M. Burke, S. Fahlvik, C. Thelander, H. Linke, M. Leijnse, Thermoelectric characterization of the Kondo resonance in nanowire quantum dots. Phys. Rev. Lett. **121** (2018)
76. B. Dutta, D. Majidi, A. Garca Corral, P.A. Erdman, S. Florens, T.A. Costi, H. Courtois, C.B. Winkelmann, Direct probe of the Seebeck coefficient in a Kondo-correlated single-quantum-dot transistor. Nano Lett. **19**, 506–511 (2019)

77. V.M. Pereira, F. Guinea, J. M. B. Lopes dos Santos, N.M.R. Peres, A.H. Castro Neto, Disorder induced localized states in graphene. Phys. Rev. Lett. **96**, 036801 (2006)
78. V.M. Pereira, J.M.B. Lopes dos Santos, A.H. Castro Neto, Modeling disorder in graphene. Phys. Rev. B **77**, 115109 (2008)
79. B. Uchoa, V.N. Kotov, N.M.R. Peres, A.H. Castro Neto, Localized magnetic states in graphene. Phys. Rev. Lett. **101**, 026805 (2008)
80. B. Uchoa, L. Yang, S.-W. Tsai, N.M.R. Peres, A.H. Castro Neto, Theory of scanning tunneling spectroscopy of magnetic adatoms in graphene. Phys. Rev. Lett. **103**, 206804 (2009)
81. K. Saha, I. Paul, K. Sengupta, Unconventional scanning tunneling conductance spectra for graphene. Phys. Rev. B **81**, 165446 (2010)
82. B. Uchoa, T.G. Rappoport, A.H. Castro Neto, Kondo quantum criticality of magnetic adatoms in graphene. Phys. Rev. Lett. **106**, 016801 (2011)
83. T.O. Wehling, H.P. Dahal, A.I. Lichtenstein, M.I. Katsnelson, H.C. Manoharan, A.V. Balatsky, Theory of Fano resonances in graphene: The influence of orbital and structural symmetries on STM spectra. Phys. Rev. B **81**, 085413 (2010)
84. O.V. Yazyev, L. Helm, Defect-induced magnetism in graphene. Phys. Rev. B **75**, 125408 (2007)
85. H. Sevinçli, M. Topsakal, E. Durgun, S. Ciraci, Electronic and magnetic properties of 3*d* transition-metal atom adsorbed graphene and graphene nanoribbons. Phys. Rev. B **77**, 195434 (2008)
86. Y. Mao, J. Yuan, J. Zhong, Density functional calculation of transition metal adatom adsorption on graphene. J. Condens. Matter Phys. **20**, 115209 (2008)
87. A.V. Krasheninnikov, P.O. Lehtinen, A.S. Foster, P. Pyykkö, R.M. Nieminen, Embedding transition-metal atoms in graphene: Structure, bonding, and magnetism. Phys. Rev. Lett. **102**, 126807 (2009)
88. H. Johll, H.C. Kang, E.S. Tok, Density functional theory study of Fe Co, and Ni adatoms and dimers adsorbed on graphene. Phys. Rev. B **79**, 245416 (2009)
89. E. Santos, A. Ayuela, D. Sánchez-Portal, First-principles study of substitutional metal impurities in graphene: structural, electronic and magnetic properties. New J. Phys. **12**, 053012 (2010)
90. M.P. Lima, A. da Silva, A. Fazzio, Adatoms in graphene as a source of current polarization: Role of the local magnetic moment. Phys. Rev. B **84**, 245411 (2011)
91. B. Nanda, M. Sherafati, Z.S. Popović, S. Satpathy, Corrigendum: Electronic structure of the substitutional vacancy in graphene: density-functional and Green's function studies. New J. Phys. **15**, 039501 (2013)
92. A. Lopez-Bezanilla, J.L. Lado, Defect-induced magnetism and Yu-Shiba-Rusinov states in twisted bilayer graphene. Phys. Rev. Mater. **3**, 084003 (2019)
93. Z.-G. Zhu, J. Berakdar, Topological view on magnetic adatoms in graphene. Phys. Rev. B **83**, 195404 (2011)
94. N.M.R. Peres, F. Guinea, A.H. Castro Neto, Electronic properties of disordered two-dimensional carbon. Phys. Rev. B **73**, 125411 (2006)
95. T.O. Wehling, A.V. Balatsky, M.I. Katsnelson, A.I. Lichtenstein, K. Scharnberg, R. Wiesendanger, Local electronic signatures of impurity states in graphene. Phys. Rev. B **75**, 125425 (2007)
96. S. Irmer, D. Kochan, J. Lee, J. Fabian, Resonant scattering due to adatoms in graphene: Top, bridge, and hollow positions. Phys. Rev. B **97**, 075417 (2018)
97. M.T. Glossop, G.E. Jones, D.E. Logan, Local quantum critical point in the pseudogap Anderson model: Finite-T dynamics and ω/T scaling. J. Phys. Chem. B **109**, 6564–72, 05 (2005)
98. L. Dell'Anna, Kondo impurity on the honeycomb lattice at half-filling. JSTAT **2010**, P01007 (2010)
99. Z.-G. Zhu, K.-H. Ding, J. Berakdar, Single- or multi-flavor Kondo effect in graphene. EPL **90**(6), 67001 (2010)

100. Z.-G. Zhu, J. Berakdar, Magnetic adatoms on graphene in the Kondo regime: An Anderson model treatment. Phys. Rev. B **84**, 165105 (2011)
101. M. Cheng, K. Ingersent, Quantum phase transitions in a pseudogap Anderson-Holstein model. Phys. Rev. B **87**, 075145 (2013)
102. T.-P. Wu, X.-Q. Wang, G.-Y. Guo, F. Anders, C.-H. Chung, Quantum criticality of the two-channel pseudogap Anderson model: universal scaling in linear and non-linear conductance. J. Phys.: Condens. Matter. **28**, 175003 (2016)
103. L. Li, Y.-Y. Ni, Y. Zhong, T.-F. Fang, H.-G. Luo, The Kondo effect of an adatom in graphene and its scanning tunneling spectroscopy. New J. Phys. **15**, 053018 (2013)
104. T. Yanagisawa, Kondo effect in Dirac systems. J. Phys. Soc. Jpn. **84**(7), 074705 (2015)
105. D. Withoff, E. Fradkin, Phase transitions in gapless Fermi systems with magnetic impurities. Phys. Rev. Lett. **64**, 1835–1838 (1990)
106. C. Gonzalez-Buxton, K. Ingersent, Renormalization-group study of Anderson and Kondo impurities in gapless Fermi systems. Phys. Rev. B **57**, 14254–14293 (1998)
107. C. Gonzalez-Buxton, K. Ingersent, Stabilization of local moments in gapless Fermi systems. Phys. Rev. B **54**, R15614–R15617 (1996)
108. T. Kanao, H. Matsuura, M. Ogata, Theory of defect-induced Kondo effect in graphene: Numerical renormalization group study. J. Phys. Soc. Jpn. **81**(6), 063709 (2012)
109. P.-W. Lo, G.-Y. Guo, F.B. Anders, Gate-tunable Kondo resistivity and dephasing rate in graphene studied by numerical renormalization group calculations. Phys. Rev. B **89**, 195424 (2014)
110. D.A. Ruiz-Tijerina, L. Dias da Silva, Transport signatures of Kondo physics and quantum criticality in graphene with magnetic impurities. Phys. Rev. B **95**, 115408 (2017)
111. G.S. Diniz, G.I. Luiz, A. Latgé, E. Vernek, From Kondo to local singlet state in graphene nanoribbons with magnetic impurities. Phys. Rev. B **97**, 115444 (2018)
112. G.-Y. Li, T.-F. Fang, A.-M. Guo, Q.-F. Sun, Ferromagnetism-induced Kondo effect in graphene with a magnetic impurity. Phys. Rev. B **100**, 115115 (2019)
113. D. Zhai, K. Ingersent, S.E. Ulloa, N. Sandler, Sublattice symmetry breaking and Kondo-effect enhancement in strained graphene. Phys. Rev. B **99**, 195410 (2019)
114. G. Diniz, G.S. Diniz, G.B. Martins, E. Vernek, Reentrant Kondo effect for a quantum impurity coupled to a metal-semiconductor hybrid contact. Phys. Rev. B **101**, 125115 (2020)
115. Y. Chang, J. Yi, A.-K. Wu, F.B. Kugler, E.Y. Andrei, D. Vanderbilt, G. Kotliar, J.H. Pixley, Vacancy-induced tunable Kondo effect in twisted bilayer graphene. Phys. Rev. Lett. **133**, 126503 (2024)
116. D. Jacob, G. Kotliar, Orbital selective and tunable Kondo effect of magnetic adatoms on graphene: Correlated electronic structure calculations. Phys. Rev. B **82**, 085423 (2010)
117. T.O. Wehling, A.V. Balatsky, M.I. Katsnelson, A.I. Lichtenstein, A. Rosch, Orbitally controlled Kondo effect of Co adatoms on graphene. Phys. Rev. B **81**, 115427 (2010)
118. D. Krychowski, J. Kaczkowski, S. Lipinski, Kondo effect of a cobalt adatom on a zigzag graphene nanoribbon. Phys. Rev. B **89**, 035424 (2014)
119. L. Liang, E. Costa Giro, and V. Meunier, "Modeling the Kondo effect of a magnetic atom adsorbed on graphene," *2D Mater.*, vol. 6, p. 035038, June 2019
120. P. Haase, S. Fuchs, T. Pruschke, H. Ochoa, F. Guinea, Magnetic moments and Kondo effect near vacancies and resonant scatterers in graphene. Phys. Rev. B **83**, 241408 (2011)
121. D. May, P.-W. Lo, K. Deltenre, A. Henke, J. Mao, Y. Jiang, G. Li, E.Y. Andrei, G.-Y. Guo, F.B. Anders, Modeling of the gate-controlled Kondo effect at carbon point defects in graphene. Phys. Rev. B **97**, 155419 (2018)
122. M. Vojta, L. Fritz, R. Bulla, Gate-controlled Kondo screening in graphene: Quantum criticality and electron-hole asymmetry. EPL **90**, 27006 (2010)
123. J.-H. Chen, L. Li, W. G. Cullen, E. D. Williams, and M. S. Fuhrer, "Tunable Kondo effect in graphene with defects," *Nat. Phys.*, vol. 7, p. 535, 04 2011
124. J. Ren, H. Guo, J. Pan, Y.Y. Zhang, X. Wu, H.-G. Luo, S. Du, S.T. Pantelides, H.-J. Gao, Kondo Effect of Cobalt Adatoms on a Graphene Monolayer Controlled by Substrate-Induced Ripples. Nano Lett. **14**, 4011 (2014)

125. Y. Jiang, P.-W. Lo, D. May, G. Li, G.-Y. Guo, F.B. Anders, T. Taniguchi, K. Watanabe, J. Mao, E.Y. Andrei, Inducing Kondo screening of vacancy magnetic moments in graphene with gating and local curvature. Nat. Commun. **9**, 2349 (2018)
126. R.R. Nair, M. Sepioni, I.-L. Tsai, O. Lehtinen, J. Keinonen, A.V. Krasheninnikov, T. Thomson, A.K. Geim, I.V. Grigorieva, Spin-half paramagnetism in graphene induced by point defects. Nature Phys. **8**, 199–202 (2012)
127. R. Nair, I.-L. Tsai, M. Sepioni, O. Lehtinen, J. Keinonen, A. Krasheninnikov, A. Castro Neto, M. Katsnelson, A. Geim, and I. Grigorieva, "Dual origin of defect magnetism in graphene and its reversible switching by molecular doping," *Nat. Commun.*, vol. 4, June 2013
128. F. Donati, Q. Dubout, G. Autès, F. Patthey, F. Calleja, P. Gambardella, O.V. Yazyev, H. Brune, Magnetic moment and anisotropy of individual Co atoms on graphene. Phys. Rev. Lett. **111**, 236801 (2013)
129. T. Eelbo, M. Waśniowska, P. Thakur, M. Gyamfi, B. Sachs, T.O. Wehling, S. Forti, U. Starke, C. Tieg, A.I. Lichtenstein, R. Wiesendanger, Adatoms and clusters of $3d$ transition metals on graphene: Electronic and magnetic configurations. Phys. Rev. Lett. **110**, 136804 (2013)
130. Y. Zhang, S.-Y. Li, H. Huang, W.-T. Li, J.-B. Qiao, W.-X. Wang, L.-J. Yin, K.-K. Bai, W. Duan, L. He, Scanning tunneling microscopy of the π magnetism of a single carbon vacancy in graphene. Phys. Rev. Lett. **117**, 166801 (2016)
131. J. Mao, Y. Jiang, D. Moldovan, G. Li, K. Watanabe, T. Taniguchi, M.R. Masir, F.M. Peeters, E.Y. Andrei, Realization of a tunable artificial atom at a supercritically charged vacancy in graphene. Nature Phys. **12**, 545–549 (2016)
132. S.-Y. Li, Y.-N. Ren, Y.-W. Liu, M.-X. Chen, H. Jiang, and L. He, "Nanoscale detection of valley-dependent spin splitting around atomic defects of graphene," *2D Materials*, vol. 6, p. 031005, jun 2019
133. Y. Zhang, F. Gao, S. Gao, L. He, Tunable magnetism of a single-carbon vacancy in graphene. Sci. Bull. **65**, 194–200 (2020)
134. P.-C. Lin, R. Villarreal, S. Achilli, H. Bana, M.N. Nair, A. Tejeda, K. Verguts, S. De Gendt, M. Auge, H. Hofsss, S. De Feyter, G. Di Santo, L. Petaccia, S. Brems, G. Fratesi, L. Pereira, Doping graphene with substitutional Mn. ACS Nano **15**, 5449–5458 (2021)
135. H. Zhang, Z. Liao, B. Xia, T.S. Herng, J. Ding, D. Gao, Ferromagnetism of $mn - n_4$ architecture embedded graphene. J. Phys. D Appl. Phys. **55**, 225001 (2022)
136. Y. Zhang, F. Gao, S. Gao, M. Brandbyge, L. He, Characterization and manipulation of intervalley scattering induced by an individual monovacancy in graphene. Phys. Rev. Lett. **129**, 096402 (2022)
137. J. Güttinger, C. Stampfer, T. Frey, T. Ihn, K. Ensslin, Graphene quantum dots in perpendicular magnetic fields. Phys. Status Solidi B **246**(11–12), 2553–2557 (2009)
138. J. Güttinger, T. Frey, C. Stampfer, T. Ihn, K. Ensslin, Spin states in graphene quantum dots. Phys. Rev. Lett. **105**, 116801 (2010)
139. J. Güttinger, C. Stampfer, T. Frey, T. Ihn, K. Ensslin, Transport through a strongly coupled graphene quantum dot in perpendicular magnetic field. Nanoscale Res. Lett. **6**, 253 (2011)
140. J. Gttinger, F. Molitor, C. Stampfer, S. Schnez, A. Jacobsen, S. Drscher, T. Ihn, K. Ensslin, Transport through graphene quantum dots. Rep. Prog. Phys. **75**, 126502 (2012)
141. T. Mller, J. Gttinger, D. Bischoff, S. Hellmller, K. Ensslin, T. Ihn, Fast detection of single-charge tunneling to a graphene quantum dot in a multi-level regime. App. Phys. Lett. **101**(1), 012104 (2012)
142. A. Jacobsen, P. Simonet, K. Ensslin, T. Ihn, Transport in a three-terminal graphene quantum dot in the multi-level regime. New J. Phys. **14**, 023052 (2012)
143. A. Jacobsen, P. Simonet, K. Ensslin, T. Ihn, Finite-bias spectroscopy of a three-terminal graphene quantum dot in the multilevel regime. Phys. Rev. B **89**, 165413 (2014)
144. M. Eich, R. Pisoni, H. Overweg, A. Kurzmann, Y. Lee, P. Rickhaus, T. Ihn, K. Ensslin, F. c. v. Herman, M. Sigrist, K. Watanabe, and T. Taniguchi, "Spin and valley states in gate-defined bilayer graphene quantum dots," *Phys. Rev. X*, vol. 8, p. 031023, Jul 2018
145. H. Overweg, H. Eggimann, X. Chen, S. Slizovskiy, M. Eich, R. Pisoni, Y. Lee, P. Rickhaus, K. Watanabe, T. Taniguchi, V. Fal'ko, T. Ihn, K. Ensslin, Electrostatically induced quantum point contacts in bilayer graphene. Nano Lett. **18**(1), 553–559 (2018)

146. F. Molitor, S. Drscher, J. Gttinger, A. Jacobsen, C. Stampfer, T. Ihn, K. Ensslin, Transport through graphene double dots. App. Phys. Lett. **94**(22), 222107 (2009)
147. F. Molitor, H. Knowles, S. Drscher, U. Gasser, T. Choi, P. Roulleau, J. Gttinger, A. Jacobsen, C. Stampfer, K. Ensslin, T. Ihn, Observation of excited states in a graphene double quantum dot. EPL **89**, 67005 (2010)
148. D. Bischoff, A. Varlet, P. Simonet, T. Ihn, K. Ensslin, Electronic triple-dot transport through a bilayer graphene island with ultrasmall constrictions. New J. Phys. **15**, 083029 (2013)
149. Q.-F. Sun, B.-G. Wang, J. Wang, T.-H. Lin, Electron transport through a mesoscopic hybrid multiterminal resonant-tunneling system. Phys. Rev. B **61**, 4754–4761 (2000)
150. G. Michałek, B.R. Bułka, T. Domański, K.I. Wysokiński, Interplay between direct and crossed Andreev reflections in hybrid nanostructures. Phys. Rev. B **88**, 155425 (2013)
151. L. Máthé, I. Grosu, Nonequilibrium Kondo effect in a graphene-coupled quantum dot in the presence of a magnetic field. Beilstein J. Nanotechnol. **11**, 225–239 (2020)
152. R. Van Roermund, *Theoretical Study of non-equilibrium transport in Kondo quantum dots* (Université de Grenoble, Theses, 2010)
153. M. Krawiec, K. Wysokiński, Charge on the quantum dot in the presence of tunneling current. Solid State Commun. **115**(3), 141–144 (2000)
154. T.-K. Ng, ac Response in the nonequilibrium Anderson impurity model. Phys. Rev. Lett. **76**, 487–490 (1996)
155. R. Fazio, R. Raimondi, Resonant andreev tunneling in strongly interacting quantum dots. Phys. Rev. Lett. **80**, 2913–2916 (1998)
156. C.A. Balseiro, G. Usaj, M.J. Snchez, Out of equilibrium transport through an Anderson impurity: probing scaling laws within the equation of motion approach. J. Phys.: Condens. Matter **22**, 425602 (2010)
157. M. Crisan, I. Grosu, Orbital Fano-Kondo effect in double quantum dots. Physica E **42**(9), 2446–2450 (2010)
158. R.M. Corless, G.H. Gonnet, D. Hare, D.J. Jeffrey, D.E. Knuth, On the Lambert$W$ function. Adv. Comput. Math. **5**, 329 (1996)
159. R. Hanson, B. Witkamp, L. Vandersypen, L. van Beveren, J.M. Elzerman, L.P. Kouwenhoven, Zeeman energy and spin relaxation in a one-electron quantum dot. Phys. Rev. Lett. **91**, 196802 (2003)
160. G. Allison, T. Fujita, K. Morimoto, S. Teraoka, M. Larsson, H. Kiyama, A. Oiwa, S. Haffouz, D.G. Austing, A. Ludwig, A.D. Wieck, S. Tarucha, Tuning the electrically evaluated electron Landé g factor in GaAs quantum dots and quantum wells of different well widths. Phys. Rev. B **90**, 235310 (2014)
161. J.N. Fuchs, M. Goerbig, Introduction to the physical properties of graphene, in *Lecture notes* (2008)
162. Y. Utsumi, J. Martinek, G. Schön, H. Imamura, S. Maekawa, Nonequilibrium Kondo effect in a quantum dot coupled to ferromagnetic leads. Phys. Rev. B **71**, 245116 (2005)
163. Žitko, J.S. Lim, R. López, R. Aguado, Shiba states and zero-bias anomalies in the hybrid normal-superconductor Anderson model. Phys. Rev. B **91**, 045441 (2015)
164. P.W. Anderson, A poor man's derivation of scaling laws for the Kondo problem. J. Phys. C: Solid State Phys. **3**(12), 2436 (1970)
165. F.D.M. Haldane, Scaling theory of the asymmetric Anderson model. Phys. Rev. Lett. **40**(6), 416–419 (1978)
166. A. Rosch, J. Paaske, J. Kroha, P. Wölfle, Nonequilibrium transport through a Kondo dot in a magnetic field: Perturbation theory and poor man's scaling. Phys. Rev. Lett. **90**(7), 076804 (2003)
167. P.B. Wiegmann, A.M. Tsvelick, Exact solution of the Anderson model: I. J. Phys. C: Solid State Phys. **16**(12), 2281 (1983)
168. A.C. Hewson, *The Kondo Problem to Heavy Fermions* (Cambridge University Press, 1993)
169. I. Affleck, A.W.W. Ludwig, The Kondo effect, conformal field theory and fusion rules. Nucl. Phys. B. **352**(3), 849–862 (1991)

Chapter 4
Optical Properties of Spherical Quantum Dots: Application of the Inversely Quadratic Hellmann Potential

This chapter of the book focuses on the theoretical study of linear and nonlinear optical properties of a spherical semiconductor quantum dot (QD). The quantum confinement for electrons in the dot is approximated using an inversely quadratic Hellmann (IQH) potential within the effective-mass approximation. The eigenenergies and corresponding eigenfunctions of the radial Schrödinger equation are calculated employing the parametric Nikiforov-Uvarov (NU) method. The optical properties of the two-level QD system, namely the optical absorption coefficient and refractive index changes, are discussed. The proposed setup could be further utilized to boost the development of novel QD-based optoelectronic devices.

4.1 State-of-The-Art and Motivation

In recent years, advances in nanofabrication technology have allowed the experimental realization of low-dimensional semiconductor heterostructures, such as QDs, quantum wires and quantum wells [1–6]. These nanostructures have also attracted the attention of researchers working in the area of solid state physics because of their unique optical and electronic properties. The presence of quantum confinement in low-dimensional nanostructures could produce nonlinear optical effects, which are relevant in the development of optoelectronic devices. The results of fundamental research on the physical properties of semiconductor materials have contributed to their wide range of applications, such as to the development of chemical and biosensors [7], optoelectronics [8], solar cells [9], photocatalysis technology [10], light-emitting diodes [10, 11] as well as biological applications [12].

As mentioned previously in Sect. 1.4, the main optical properties of low-dimensional semiconductor nanojunctions are regulated by the optical intersubband transitions. Namely, when the energy of incident photon equals the energy corresponding to the optical intersubband transition, then the dielectric constant of the

© The Author(s), under exclusive license to Springer Nature Switzerland AG 2025
L. Máthé, *Physical Properties of Low-Dimensional Systems*,
NanoScience and Technology, https://doi.org/10.1007/978-3-031-91602-1_4

irradiated material can exhibit significant changes leading to a modification in its optical absorption coefficient, and refractive index. The optical intersubband transition is characterized by the matrix elements of electric dipole moment which highly affect the optical properties of the low-dimensional nanostructures. It has been found that the nonlinear contributions of optical properties of QDs can be amplified by simultaneously increasing the values of matrix element of electric dipole moment and decreasing the energy gap corresponding to the subbands [13, 14]. The theoretical studies in the literature indicate that the shape of spatial confinement potential confining the motion of charge carriers in QDs plays an important role in the determination of optical properties of QDs [13–15]. Up to now, a plethora of potential models, such as Woods-Saxon [16, 17], Gaussian [18], modified Gaussian [19, 20], Morse [21], Rosen-Morse [22, 23], parabolic [24–27], cylindrical [28], Tietz [29], modified Pöschl-Teller [30], Kratzer [31, 32], modified Kratzer [33], ring-shaped non-spherical oscillator [34], Hulthén [35], Hulthén-Yukawa [36], Manning-Rosen [37], Deng-Fan-Eckart [38], radial potential [15], Konwent-like potential [39] have been proposed in order to determine the optical properties of QDs. Besides these theoretical studies, the optical properties of low-dimensional semiconductor nanostructures have also been investigated experimentally [40–46].

The aim of this chapter is to theoretically investigate the optical properties of a spherical QD in which the spatial confinement of charge carriers is realized by the IQH potential. For this purpose, we give analytical formulas for the linear and third-order nonlinear terms of optical absorption coefficient and corresponding changes in refractive index for a two-level quantum system based on the density matrix approach by using perturbation theory [26, 47, 48]. The time-independent Schrödinger equation describing the QD electron is separated into a radial and orbital part. We solve the radial Schrödinger equation by using the parametric NU method [49, 50]. We discuss in details the influence of the QD characteristic size, confinement potential depth, and incident optical intensity on the absorption coefficient and refractive index changes in the case of a two-level QD system. We found that the position of resonances (or antiresonance) of optical absorption coefficients and intersection point of refractive index changes shift when tuning the potential depth or changing the QD characteristic size. In addition, the results show that the increasing incident optical intensity suppresses the magnitude of optical absorption coefficient and refractive index changes.

4.2 Theoretical Model

4.2.1 Model

As discussed previously in Sect. 1.1, the spatial confinement of charge carriers leads to the emergence of discrete energy levels in the spectrum of the QD. The confinement induced energy levels in low-dimensional semiconductor heterostructures, such as in QDs, can be modified by changing the potential characteristics via voltages applied to the gate electrodes placed near the dot. In our model, we approximate the spatial confinement for the motion of an electron in a spherically symmetric QD with an IQH potential, which is expressed as [51–54]:

$$V(r) = -\frac{\mathcal{A}}{r} + \frac{\mathcal{B}}{r^2}e^{-\alpha r}, \tag{4.1}$$

where α is a screening parameter through which we define a QD characteristic size as $R_0 = 1/\alpha$. Here, the parameters $\mathcal{A}$ and $\mathcal{B}$ regulate the depth of confinement potential via the definition $\mathcal{A} = V_0 R_0$ and $\mathcal{B} = V_0 R_0^2$ with $V_0 > 0$. We consider small values of αr in $V(r)$ given by the relation (4.1), thus we can expand $\exp(-\alpha r)$ by Taylor series up to term including $\mathcal{O}((\alpha r)^2)$. Equation (4.1) reduces to

$$V(r) \approx V_0\left[\left(\frac{R_0}{r}\right)^2 - 2\frac{R_0}{r}\right], \tag{4.2}$$

where we added a $-V_0/2$ constant to the expanded form of the IQH potential in order to ensure that $\lim_{r\to\infty} V(r) = 0$, according to Eq. (4.1). Note that the second-order Taylor series expansion of the IQH potential (4.1), in terms of the dimensionless distance αr, reproduces the Kratzer [31, 32] or a modified Kratzer-type potential [33], up to an additive constant.

Let us consider an electron in a spherical QD with the IQH potential. Within the framework of effective-mass approximation, the time-independent Schrödinger equation of the system, in spherical coordinates, is written as [55]

$$\begin{aligned} -\frac{\hbar^2}{2m_e^*}\Bigg[&\frac{1}{r^2}\frac{\partial}{\partial r}\left(r^2\frac{\partial}{\partial r}\right) + \frac{1}{r^2\sin\theta}\frac{\partial}{\partial\theta}\left(\sin\theta\frac{\partial}{\partial\theta}\right) \\ &+ \frac{1}{r^2\sin^2\theta}\frac{\partial^2}{\partial\phi^2}\Bigg]\psi_{n\ell m}(r,\theta,\phi) = [E_{n\ell} - V(r)]\psi_{n\ell m}(r,\theta,\phi), \end{aligned} \tag{4.3}$$

where m_e^* and $V(r)$ denote the effective mass of an electron and the spherically symmetric confinement potential, respectively. Here, $E_{n\ell}$ and $\psi_{n\ell m}(r,\theta,\phi)$ represent the eigenvalues and eigenfunctions of Eq. (4.3), with n, ℓ and m being the principal, orbital and magnetic quantum numbers. We recall that for bound states the eigenvalues are negative real numbers, $E_{n\ell} < 0$. The variables θ and ϕ denote the polar and azimuthal angles, respectively. Since the differential equation in relation (4.3) is separable, the solutions are of the form:

$$\psi_{n\ell m}(r,\theta,\phi) = \mathcal{R}_{n\ell}(r)Y_\ell^m(\theta,\phi), \tag{4.4}$$

where $\mathcal{R}_{n\ell}(r)$ and $Y_\ell^m(\theta,\phi)$ are called the radial wavefunction and the spherical harmonic functions. Substituting the form of $\psi_{n\ell m}(r,\theta,\phi)$ from relation (4.4) into the Schrödinger equation given by Eq. (4.3), one obtains

$$\frac{d^2\mathcal{R}_{n\ell}(r)}{dr^2} + \frac{2}{r}\frac{d\mathcal{R}_{n\ell}(r)}{dr} + \frac{2m_e^*}{\hbar^2}\left[E_{n\ell} - V(r) - \frac{\ell(\ell+1)\hbar^2}{2m_e^* r^2}\right]\mathcal{R}_{n\ell}(r) = 0, \tag{4.5}$$

being the radial Schrödinger equation, and

$$-\left\{\frac{\hbar^2}{\sin^2\theta}\left[\sin\theta\frac{\partial}{\partial\theta}\left(\sin\theta\frac{\partial}{\partial\theta}\right)+\frac{\partial^2}{\partial\phi^2}\right]\right\}Y_\ell^m(\theta,\phi)=\ell(\ell+1)\hbar^2\,Y_\ell^m(\theta,\phi), \tag{4.6}$$

standing for the square of orbital angular momentum, $\boldsymbol{L}^2$, with the spherical harmonics $Y_\ell^m(\theta,\phi)$ eigenfunctions [56]. Substituting the approximated expression of the IQH potential from Eq. (4.2) into Eq. (4.5), we find

$$\begin{aligned}\frac{d^2\mathcal{R}_{n\ell}(r)}{dr^2}+\frac{2}{r}\frac{d\mathcal{R}_{n\ell}(r)}{dr}+\frac{2m_e^*}{\hbar^2}\Bigg[-\left(\frac{\ell(\ell+1)\hbar^2}{2m_e^*}+V_0R_0^2\right)\frac{1}{r^2}\\+2V_0R_0\frac{1}{r}+E_{n\ell}\Bigg]\mathcal{R}_{n\ell}(r)=0.\end{aligned} \tag{4.7}$$

Equation (4.7) is a second-order ordinary differential equation that we solve by using the parametric NU method [49, 50]. The NU method allows us to solve hypergeometric-type second-order linear differential equations by employing special orthogonal functions. Such Schrödinger-type equations expressed in spherical coordinates with a given potential are transformed into generalized hypergeometric-type equations allowing thus the application of NU method to find their exact or particular eigensolutions [55]. The detailed determination of solutions of Eq. (4.7) can be found in Appendix of this chapter. The eigenenergy solutions thus read

$$E_{n\ell}=-\frac{8m_e^*}{\hbar^2}\left(\frac{V_0R_0}{(2n+1)+\sqrt{(2\ell+1)^2+\frac{8m_e^*}{\hbar^2}V_0R_0^2}}\right)^2, \tag{4.8}$$

with the corresponding eigenfunctions,

$$\mathcal{R}_{n\ell}(r)=\mathcal{N}_{n\ell}\,r^{\frac{1}{2}(\mathcal{Z}_\ell-1)}e^{-\mathcal{K}_{n\ell}r}\mathcal{L}_n^{\mathcal{Z}_\ell}(2\mathcal{K}_{n\ell}r), \tag{4.9}$$

where $\mathcal{L}_n$ stands for the Laguerre polynomials with the following notations:

$$\begin{aligned}\mathcal{Z}_\ell&=\sqrt{(2\ell+1)^2+\frac{8m_e^*}{\hbar^2}V_0R_0^2},\\ \mathcal{K}_{n\ell}&=\sqrt{-\frac{2m_e^*}{\hbar^2}E_{n\ell}}.\end{aligned} \tag{4.10}$$

The eigenfunctions of Eq. (4.3) are obtained by substituting the solutions given by Eq. (4.9) into Eq. (4.4). The normalization constants $\mathcal{N}_{n\ell}$ are determined with the help of the relation $\int\psi_{n\ell m}(r,\theta,\phi)\psi_{n\ell m}^*(r,\theta,\phi)d\Omega=1$ where $d\Omega=r^2dr\sin\theta d\theta d\phi$.

Knowing now the energy levels of the QD, $E_{n\ell}$, with the corresponding wavefunctions $\psi_{n\ell m}(r,\theta,\phi)$, the optical properties, such as the absorption coefficients and refractive index changes, in a two-level QD system can be analyzed.

4.2.2 Optical Absorption Coefficient and Refractive Index Changes

As mentioned previously in Sect. 1.4.2, the optical absorption coefficient $\alpha(\omega)$ and refractive index changes $\Delta n_r/n_r$ in the QD system are determined by the susceptibility $\chi(\omega)$. The optical absorption coefficient can be calculated via the formula [22, 47, 48, 57, 58]:

$$\alpha(\omega) = \omega\sqrt{\frac{\mu}{\varepsilon}}\mathrm{Im}[\varepsilon_0\chi(\omega)], \tag{4.11}$$

where $\mu = \mu_r\mu_0$ stands for the system permeability with μ_r and μ_0 being the relative and vacuum permeability. $\varepsilon = n_r^2\varepsilon_0$ represents the materials' permittivity with the refractive index, n_r. Based on Eq. (1.29), Eq. (4.11) can be written as the sum of the linear $\alpha^{(1)}(\omega)$ and third-order nonlinear $\alpha^{(3)}(\omega)$ components of the total optical absorption coefficient $\alpha(\omega)$:

$$\begin{aligned}\alpha(\omega) &\approx \alpha^{(1)}(\omega) + \alpha^{(3)}(\omega)\\ &= \omega\sqrt{\frac{\mu}{\varepsilon}}\mathrm{Im}\big[\varepsilon_0\chi^{(1)}(\omega) + \varepsilon_0\chi^{(3)}(\omega)|\tilde{E}|^2\big].\end{aligned} \tag{4.12}$$

Using Eqs. (1.30)–(1.31) and (4.12), the linear and third-order nonlinear optical absorption coefficients read

$$\alpha^{(1)}(\omega) = \omega\sqrt{\frac{\mu}{\varepsilon}}\frac{\sigma_v\hbar\Gamma_{12}|\mathcal{M}_{21}|^2}{(E_{21}-\hbar\omega)^2 + (\hbar\Gamma_{12})^2}, \tag{4.13}$$

and

$$\begin{aligned}\alpha^{(3)}(\omega) = &-\omega\sqrt{\frac{\mu}{\varepsilon}}\frac{I}{2\varepsilon_0 n_r c}\frac{\sigma_v\hbar\Gamma_{12}|\mathcal{M}_{21}|^2}{[(E_{21}-\hbar\omega)^2 + (\hbar\Gamma_{12})^2]^2}\\ &\times\left\{4|\mathcal{M}_{21}|^2 - \frac{|\mathcal{M}_{22}-\mathcal{M}_{11}|^2[3E_{21}^2 - 4E_{21}\hbar\omega + \hbar^2(\omega^2-\Gamma_{12}^2)]}{E_{21}^2 + (\hbar\Gamma_{12})^2}\right\},\end{aligned} \tag{4.14}$$

respectively. Here, $I = 2\varepsilon_0 n_r c|\tilde{E}|^2$ denotes the incident optical intensity with c being the speed of light in vacuum.

The relative changes in refractive index are connected to the susceptibility via the relation [22, 48, 57, 58]:

$$\frac{\Delta n(\omega)}{n_r} = \mathrm{Re}\Big(\frac{\chi(\omega)}{2n_r^2}\Big). \tag{4.15}$$

Based on Eq. (1.29), the refractive index change $\Delta n(\omega)/n_r$ given by Eq. (4.15) is equal to the sum of its linear $\Delta n^{(1)}(\omega)/n_r$ and third-order nonlinear $\Delta n^{(3)}(\omega)/n_r$ components,

$$\begin{aligned}\frac{\Delta n(\omega)}{n_r} &\approx \frac{\Delta n^{(1)}(\omega)}{n_r} + \frac{\Delta n^{(3)}(\omega)}{n_r} \\ &= \mathrm{Re}\Big(\frac{1}{2n_r^2}\big[\chi^{(1)}(\omega) + \chi^{(3)}(\omega)|\tilde{E}|^2\big]\Big).\end{aligned} \tag{4.16}$$

Using Eqs. (1.30)-(1.31) and (4.16), the linear and third-order nonlinear refractive index changes become

$$\frac{\Delta n^{(1)}(\omega)}{n_r} = \frac{\sigma_\nu |\mathcal{M}_{21}|^2}{2n_r^2\varepsilon_0} \frac{E_{21} - \hbar\omega}{(E_{21} - \hbar\omega)^2 + (\hbar\Gamma_{12})^2}, \tag{4.17}$$

and

$$\begin{aligned}\frac{\Delta n^{(3)}(\omega)}{n_r} = -\frac{\sigma_\nu |\mathcal{M}_{21}|^2}{4n_r^3\varepsilon_0} \frac{\mu_0 c I}{\big[(E_{21} - \hbar\omega)^2 + (\hbar\Gamma_{12})^2\big]^2} \Bigg[4(E_{21} - \hbar\omega)|\mathcal{M}_{21}|^2 \\ - \frac{|\mathcal{M}_{22} - \mathcal{M}_{11}|^2}{E_{21}^2 + (\hbar\Gamma_{12})^2} \Big\{(E_{21} - \hbar\omega)\big[E_{21}(E_{21} - \hbar\omega) - (\hbar\Gamma_{12})^2\big] \\ - (\hbar\Gamma_{12})^2(2E_{21} - \hbar\omega)\Big\}\Bigg],\end{aligned} \tag{4.18}$$

respectively.

Some remarks regarding the electric dipole moment $\mathcal{M}$ are needed. Firstly, the spherical symmetry of the considered system leads to vanishing diagonal elements of electric dipole moment, i.e., $\mathcal{M}_{ii} = 0$. The non-diagonal matrix element of the electric dipole moment $\mathcal{M}_{ij}$ (where $i \neq j$) corresponds to a transition between two electronic states, typically between the ground and first excited states. The probability amplitude of such an optical transition depends on $\mathcal{M}_{ij}$. Since $\mathcal{M}$ is a vector, its direction determines the polarization of a given transition and gives the response of a system to a polarized applied electromagnetic field. Also, it expresses the manner of electric field changes caused by the material. Since the electric dipole moment does not depend on the spin, the optical transitions can take place only between states with the same spin. Thus, the matrix elements of the electric dipole moment $\mathcal{M}_{ij}$ do not vanish when the selection rules $\Delta\ell = \pm 1$ and $\Delta m = 0$ are fulfilled [59, 60]. Since the diagonal elements of $\mathcal{M}$ vanish, i.e., $\mathcal{M}_{ii} = 0$, this results in $\alpha^{(3)}(\omega)$ having negative values, and $\Delta n^{(3)}(\omega)/n_r$ taking the opposite sign to $\Delta n^{(1)}(\omega)/n_r$. Therefore, the optical absorption coefficients and relative changes in refractive index depend only on the $|\mathcal{M}_{21}|$ matrix element of the electric dipole moment $\mathcal{M}$. In addition, we observe that both $\alpha^{(3)}(\omega)$ and $\Delta n^{(3)}(\omega)/n_r$ are proportional to I.

4.3 Results and Discussion

In this section, we present our results on the optical properties of a QD realized from a GaAs heterostructure. The effective mass, carrier density, relative permeability, refractive index, and relaxation time are taken as $m_e^* = 0.067\, m_e, \sigma_v = 5 \cdot 10^{22}\text{m}^{-3}$, $\mu_r = 1, n_r = 3.2$, and $T_{12} = 0.14\,\text{ps}$ [16, 20, 23]. Here, $\varepsilon_0 = 8.854 \cdot 10^{-12}\,\text{F/m}$ and $\mu_0 = 4\pi \cdot 10^{-7}\,\text{H/m}$ represent the vacuum permittivity and vacuum permeability. The potential depth V_0 and characteristic size of QD R_0 are selected according to the literature [18, 33, 61]. We chose the ground and first excited states as: $n = 0$, $\ell = 0$ and $n = 0$, $\ell = 1$. For simplicity, in the following calculations, we use the atomic units, i.e., $\hbar = m_e = e = 1$.

We first study the behavior of the IQH potential that is used in our model for spatial confinement of electrons in the QD. We plot the approximated form of the IQH potential, $V(r)$, according to Eq. (4.2), as a function of the distance r for different values of the characteristic size of QD R_0 at a constant value of the potential depth V_0 in Fig. 4.1a and for different values of the potential depth V_0 at a fixed value of the QD characteristic size R_0 in Fig. 4.1b. We observe that the IQH potential is a continuous function in r and is characterized by the potential depth V_0 and QD characteristic size R_0. Owing to $[\partial^2 V(r)/\partial r^2]_{r=R_0} > 0$, the potential $V(r)$ shows a local minimum at distance $r = R_0$ with depth $V(r = R_0) = -V_0$. At larger distances, the potential $V(r)$ becomes independent of the QD characteristic size and it takes zero value, i.e., $\lim_{r\to\infty} V(r) = 0$.

In the following, we investigate the influence of the dot characteristic size R_0, potential depth V_0 and incident optical intensity I on the linear, third-order nonlinear and total absorption coefficients, and on the linear, third-order nonlinear and total refractive index changes, respectively. We first show in Fig. 4.2a the results connected to the behavior of linear and third-order nonlinear absorption coefficients, as well as on the total absorption coefficient, as a function of photon energy $\hbar\omega$ when the dot characteristic size R_0 is varied and the potential depth V_0 is fixed. We also

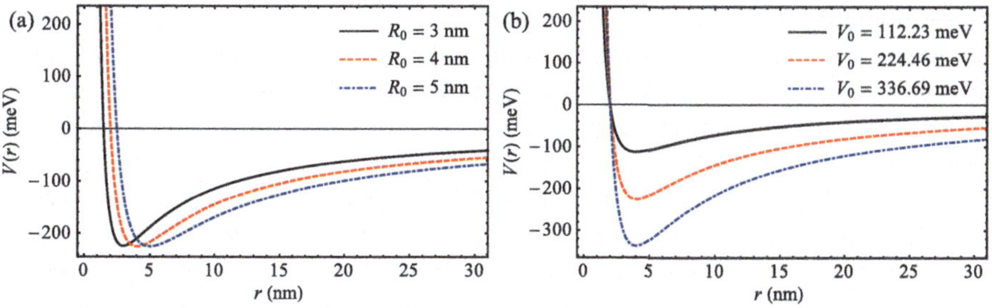

Fig. 4.1 The confinement IQH potential against the distance r: **a** at different values of the QD characteristic size R_0 with a fixed potential depth $V_0 = 224.46\,\text{meV}$, and **b** at different values of the potential depth V_0 with a fixed QD characteristic size $R_0 = 4\,\text{nm}$. Adapted with permission from [54]. Copyright 2021, Elsevier Science & Technology Journals

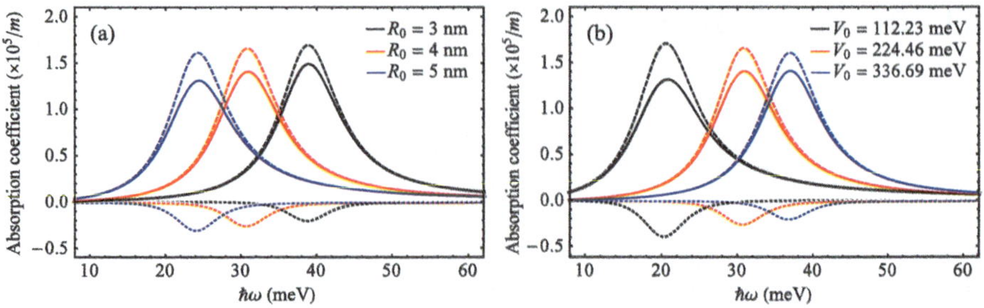

Fig. 4.2 The variation of linear [dashed line, $\alpha^{(1)}(\omega)$], third-order nonlinear [dotted line, $\alpha^{(3)}(\omega)$], and total [solid line, $\alpha(\omega)$] optical absorption coefficients with the incident photon energy $\hbar\omega$: **a** at different values of the QD characteristic size R_0 for a fixed potential depth value $V_0 = 224.46$ meV, and **b** at different values of the potential depth V_0 for a fixed value of the QD characteristic size $R_0 = 4$ nm. The optical intensity is $I = 8 \cdot 10^8$ W/m^2. Reproduced with permission from [54]. Copyright 2021, Elsevier Science & Technology Journals

plot in Fig. 4.2b the corresponding absorption coefficients when the potential depth V_0 is tuned and the dot characteristic size R_0 is fixed. The optical intensity I is taken as a constant. We can observe that the linear $\alpha^{(1)}(\omega)$ and the total $\alpha(\omega)$ optical absorption coefficients reach maximum values when the energy of the incident photon matches the energy difference between the two subbands, i.e., $\hbar\omega = E_{21}$. In contrast to $\alpha^{(1)}(\omega)$ and $\alpha(\omega)$, the third-order nonlinear optical absorption coefficient $\alpha^{(3)}$ presents a minimum at $\hbar\omega = E_{21}$. In addition, the maximum of $\alpha^{(1)}(\omega)$ and $\alpha(\omega)$ as well as the minimum of $\alpha^{(3)}(\omega)$ are shifted to the lower energies (redshifted) when the dot characteristic size R_0 is increasing [see Fig. 4.2a]. This behavior of the optical absorption coefficients can be explained with the change in energy difference E_{21} as a response to the change in R_0. Since, as the dot characteristic size R_0 increases, the energy difference E_{21} decreases, for values of R_0 considered here, resulting thus in a shift of $\alpha^{(1)}(\omega)$ and $\alpha(\omega)$ maxima (or $\alpha^{(3)}(\omega)$ minimum) to the lower values of the photon energy $\hbar\omega$. Furthermore, the QD characteristic size R_0 influences the amplitude of optical absorption coefficients via the $\mathcal{M}_{21}$ matrix element of electric dipole moment. Due to the dependence of $\alpha^{(1)}(\omega) \propto |\mathcal{M}_{21}|^2$ and $|\alpha^{(3)}(\omega)| \propto |\mathcal{M}_{21}|^4$, the effect of R_0 on the changes in the amplitude of linear and third-order nonlinear absorption coefficients manifests differently. Namely, the transition matrix element $|\mathcal{M}_{21}|$ increases with dot characteristic size R_0, for values of R_0 considered here, resulting in a more observably change in $\alpha^{(3)}(\omega)$ due to its higher-order dependence on $|\mathcal{M}_{21}|$. In the case when the potential depth V_0 is varied and the dot characteristic size R_0 is fixed [see Fig. 4.2b], the position of linear and total optical absorption coefficients maxima, or the minimum of third-order nonlinear optical absorption coefficient are shifted to the higher energy values (blueshifted) with the increase of V_0. This is due to the fact that the energy difference between subbands E_{21} increases with the magnitude of V_0. In addition, by increasing V_0 the matrix element $|\mathcal{M}_{21}|$ is starting to decrease which causes the shallowing of the $\alpha^{(3)}(\omega)$ valley.

We plot in Fig. 4.3a–c the linear $\Delta n^{(1)}(\omega)/n_r$, third-order nonlinear $\Delta n^{(3)}(\omega)/n_r$, and total $\Delta n(\omega)/n_r$ refractive index changes as a function of photon energy $\hbar\omega$ when the QD characteristic size R_0 is varied at a constant V_0. Figure 4.3d–f show the results for the corresponding changes in the refractive index when the potential depth V_0 is tuned for a given QD characteristic size R_0 value. The optical intensity I has a single value in both cases. It can be observed that the linear $\Delta n^{(1)}(\omega)/n_r$, third-order nonlinear $\Delta n^{(3)}(\omega)/n_r$ and total $\Delta n(\omega)/n_r$ refractive index changes intersect the energy-axis when the incident photon energy equals the energy difference between the levels corresponding to the first excited state and ground state, i.e., $\hbar\omega = E_{21}$. In addition, the position of intersection point shifts to lower energies (redshift) with the increase of dot characteristic size R_0 at a fixed V_0, as it is illustrated in Fig. 4.3a–c. The shift of matching point to lower energies is due to the decreasing of energy difference E_{21} with the increase of QD characteristic size R_0 values considered here. This is also

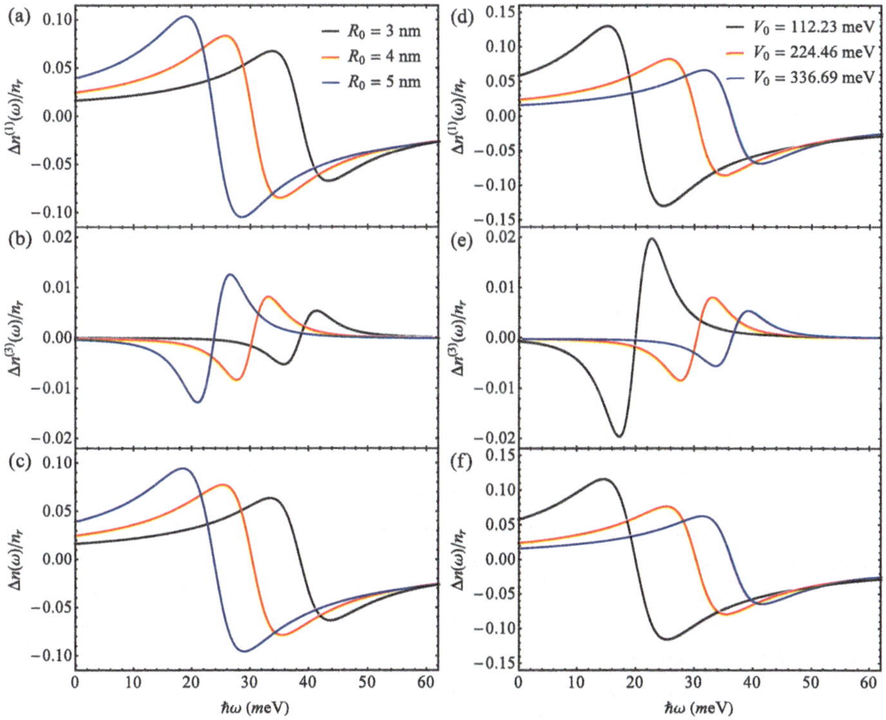

Fig. 4.3 The results for **a** linear $\Delta n^{(1)}(\omega)/n_r$, **b** third-order nonlinear $\Delta n^{(3)}(\omega)/n_r$, and **c** total $\Delta n(\omega)/n_r$ refractive index changes against the energy of incident photon $\hbar\omega$ at different values of the QD characteristic size R_0 with a fixed potential depth value $V_0 = 224.46$ meV. **d** The linear $\Delta n^{(1)}(\omega)/n_r$, **e** third-order nonlinear $\Delta n^{(3)}(\omega)/n_r$ and (f) total $\Delta n(\omega)/n_r$ refractive index changes against the energy of incident photon $\hbar\omega$ at different values of the potential depth V_0 with a fixed value of the QD characteristic size $R_0 = 4$ nm. The optical intensity is $I = 8 \cdot 10^8$ W/m^2. Reproduced with permission from [54]. Copyright 2021, Elsevier Science & Technology Journals

discussed in the context of Fig. 4.2. The amplitudes of the refractive index changes grow with the increase of dot characteristic size R_0 due to the larger matrix element $|\mathcal{M}_{21}|$. When the QD characteristic size R_0 takes a constant value and the magnitude of V_0 is varied [see Fig. 4.3d–f], the amplitude of linear $\Delta n^{(1)}(\omega)/n_r$, third-order nonlinear $\Delta n^{(3)}(\omega)/n_r$ and total $\Delta n(\omega)/n_r$ refractive index changes reduces with the increase of V_0. The reduction in amplitudes of the refractive index changes with the increase of V_0 is explained with the decreasing matrix element $|\mathcal{M}_{21}|$ when the potential depth V_0 is increasing. Furthermore, the enlarged potential depth V_0 leads to the growth of energy difference E_{21} which causes the intersection point of refractive index changes starts to shift to higher energy values (blueshift).

Finally, we study the effect of incident optical intensity I on absorption coefficient and refractive index changes. Figure 4.4a shows the total optical absorption coefficient against the incident photon energy $\hbar\omega$ at different values of the optical intensity I for a fixed value of the QD characteristic size R_0 and potential depth V_0. We observe that the amplitude of total optical absorption coefficient decreases with the increase of incident optical intensity I. This behavior of $\alpha(\omega)$ with the changes of I is explained as follows. The linear optical absorption coefficient $\alpha^{(1)}(\omega)$ does not depend on I, but the third-order nonlinear optical absorption coefficient $\alpha^{(3)}$ shows a linear dependence on the optical intensity [see Eqs. (4.12)–(4.14)]. Thus, the increase of optical intensity I results in a reduction of the magnitude of total optical absorption coefficient $\alpha(\omega)$. A further increase in optical intensity leads to the appearance of a minimum in the spectrum of $\alpha(\omega)$ at photon energy $\hbar\omega \approx E_{21}$ for the value of optical intensity $I \approx 2.65 \cdot 10^9\ \mathrm{W/m^2}$ for which the absolute value of third-order nonlinear and total optical absorption coefficients are equal, i.e., $|\alpha^{(3)}| \approx \alpha$. Further enhancement in the intensity I leads to the creation of a dip in the spectrum of total absorption coefficient and thus to the emergence of two resonant peaks. When the third-order nonlinear absorption coefficient approaches the value of linear absorption coefficient at $\hbar\omega \approx E_{21}$, i.e., $|\alpha^{(3)}| \approx \alpha^{(1)}$, further increase in I leads incorrectly to negative values of α where the Eq. (4.12) is no longer applicable [62]. This is explained with the fact that the perturbation theory used for calculating susceptibilities allows values for optical intensity I for which $|\alpha^{(3)}(\omega)|$ is much lesser than $\alpha^{(1)}(\omega)$, i.e., $|\alpha^{(3)}(\omega)| \ll \alpha^{(1)}(\omega)$. Note also that the position of resonant peak of $\alpha(\omega)$ is not affected by the optical intensity I. Figure 4.4b illustrates the total refractive index changes versus the photon energy $\hbar\omega$ at different values of the optical intensity I for fixed values of R_0 and V_0. We see from Eqs. (4.16)–(4.18) that the linear refractive index changes do not depend on the optical intensity, while the third-order nonlinear term linearly depends on I. Consequently, the total refractive index changes depend on optical intensity via $\Delta n^{(3)}(\omega)/n_r$. At low optical intensities, the contribution of $\Delta n^{(1)}(\omega)/n_r$ to the total refractive index changes is dominant, but with increasing I, $|\Delta n^{(3)}(\omega)/n_r|$ becomes larger and the amplitude of $\Delta n(\omega)/n_r$ begins to decrease. Similarly to the behavior of $\alpha(\omega)$, when the linear and third-order nonlinear refractive index changes cancel each other, a further increase in the optical intensity leads incorrectly to altering the magnitude of total refractive index changes. This is due to

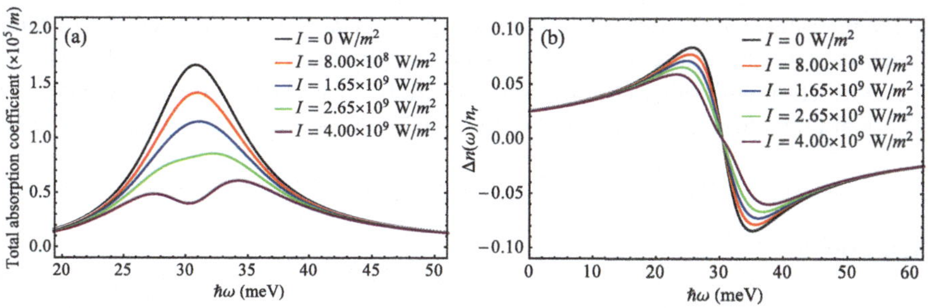

Fig. 4.4 **a** The total optical absorption coefficient $\alpha(\omega)$ and **b** total refractive index changes $\Delta n(\omega)/n_r$ against the energy of incident photon $\hbar\omega$ at different values of the optical intensity I with a fixed value of potential depth $V_0 = 224.46$ meV and QD characteristic size $R_0 = 4$ nm. Reproduced with permission from [54]. Copyright 2021, Elsevier Science & Technology Journals

the fact that for values of I where $|\Delta n^{(3)}(\omega)/n_r|$ is larger than $|\Delta n^{(1)}(\omega)/n_r|$, the approximations used in Sect. 1.4.2 do not hold anymore. The position of matching point in the spectrum of $\Delta n(\omega)/n_r$ is not affected by the optical intensity.

4.4 Summary

In this chapter, we have theoretically studied the linear and nonlinear optical properties of a spherical QD, in which the spatial confinement of electrons has been modeled with an IQH potential. The main optical properties of the system are based on the density of induced electric dipole moments that has been treated via a power series expansion in the strength of electric field. The linear, third-order nonlinear and total optical absorption coefficients and refractive index changes have been determined from the corresponding susceptibilities, calculated via density matrix approach. The time-independent Schrödinger equation in spherical coordinates, which describes the motion of an electron in QD, has been separated into a radial and an orbital equation. The eigensolutions of the radial equation have been found with the help of parametric NU method. The eigenfunctions of the orbital part of Schrödinger equation are given by the spherical harmonics. The optical properties of the QD system with IQH potential have been studied in details. It has been found that the position of resonances (or antiresonance) in the spectrum of optical absorption coefficients and intersection point of refractive index changes can be shifted by varying the QD characteristic size or the value of confinement potential depth. We have also established that the increased incident optical intensity suppresses the amplitude of absorption coefficients and refractive index changes showing no influence on the position of their corresponding resonances or matching points. Our findings are consistent with results obtained using different types of potential models proposed in the literature [16–18, 48]. Starting with our study [54], the IQH potential has been effectively applied in the literature to theoretically analyze the influence of hydrostatic pressure and temperature on the optical rectification [63], as well as on the second and third harmonic generation susceptibilities [64–66] of spherical QDs. It has also been used, in combination with other confinement potentials, to study the optical properties of spherical QDs [67–71] and to analyze the optical properties of quantum wells [72].

Appendix

As mentioned in Sect. 4.2, the NU method allows us to solve Schrödinger-like differential equations in spherical coordinates by transforming them into a generalized hypergeometric-type differential equation [49, 55]. The main second-order ordinary differential equation, which forms the basic of the NU method, can be written in the following form

$$\frac{d^2\psi(s)}{ds^2} + \frac{\tilde{\tau}(s)}{\sigma(s)}\frac{d\psi(s)}{ds} + \frac{\tilde{\sigma}(s)}{\sigma^2(s)}\psi(s) = 0, \tag{4.19}$$

where $\psi(s)$ being a function of hypergeometric-type and $\tilde{\sigma}(s)$ as well as $\sigma(s)$ represent polynomials at most second-degree, while $\tilde{\tau}(s)$ stands for a polynomial at most of first-degree. Based on Ref. [50], Eq. (4.19) can be written in a parametric form of

$$\frac{d^2\psi(s)}{ds^2} + \frac{c_1 - c_2 s}{s(1 - c_3 s)}\frac{d\psi(s)}{ds} + \frac{-\xi_1 s^2 + \xi_2 s - \xi_3}{s^2(1 - c_3 s)^2}\psi(s) = 0, \tag{4.20}$$

which eigenfunctions are given by

$$\psi(s) = \mathcal{N}_{nl}\, s^{c_{12}}\, (1 - c_3 s)^{-c_{12} - \frac{c_{13}}{c_3}}\, \mathcal{P}_n^{\left(c_{10}-1,\, \frac{c_{11}}{c_3} - c_{10} - 1\right)}(1 - 2c_3 s), \tag{4.21}$$

where $\mathcal{P}_n^{(\alpha,\beta)}(x)$ are the Jacobi polynomials. In the $c_3 = 0$ limit, Eq. (4.21) reduces to

$$\psi(s) = \mathcal{N}_{nl}\, s^{c_{12}}\, e^{c_{13} s}\, \mathcal{L}_n^{c_{10}-1}(c_{11} s), \tag{4.22}$$

a simpler form, where $\mathcal{L}_n^{(\alpha)}(x)$ represents the associated Laguerre polynomials. The energy eigenvalues are obtained from the following equation

$$c_3 n^2 + (c_2 - c_3)n - (2n + 1)c_5 + (2n + 1)(\sqrt{c_9} + c_3\sqrt{c_8}) + c_7 + 2\, c_3\, c_8 + 2\sqrt{c_8 c_9} = 0, \tag{4.23}$$

with parameters

$$c_4 = \frac{1}{2}(1 - c_1),\quad c_5 = \frac{1}{2}(c_2 - 2c_3),\quad c_6 = c_5^2 + \xi_1,\quad c_7 = 2\, c_4\, c_5 - \xi_2, \tag{4.24}$$
$$c_8 = c_4^2 + \xi_3,\quad c_9 = c_3\, c_7 + c_3^2\, c_8 + c_6,\quad c_{10} = c_1 + 2\, c_4 + 2\sqrt{c_8},$$
$$c_{11} = c_2 - 2\, c_5 + 2(\sqrt{c_9} + c_3\sqrt{c_8}),\quad c_{12} = c_4 + \sqrt{c_8},\quad c_{13} = c_5 - (\sqrt{c_9} + c_3\sqrt{c_8}).$$

Comparing now Eq. (4.7) with Eq. (4.20), one finds

$$
\begin{aligned}
c_1 &= 2, \quad c_2 = c_3 = 0, \\
\xi_1 &= -\frac{2m_e^*}{\hbar^2} E_{nl} \\
\xi_2 &= \frac{4m_e^*}{\hbar^2} V_0 R_0 \\
\xi_3 &= \frac{2m_e^*}{\hbar^2}\left(\frac{\ell(\ell+1)\hbar^2}{2m_e^*} + V_0 R_0^2\right).
\end{aligned}
\tag{4.25}
$$

Using Eq. (4.25), the values of parameters, given in Eq. (4.24), read

$$
\begin{aligned}
& c_4 = -\frac{1}{2}, \quad c_5 = 0, \quad c_6 = -\frac{2m_e^*}{\hbar^2} E_{nl}, \quad c_7 = -\frac{4m_e^*}{\hbar^2} V_0 R_0, \\
& c_8 = \frac{2m_e^*}{\hbar^2}\left(\frac{\ell(\ell+1)\hbar^2}{2m_e^*} + V_0 R_0^2\right) + \frac{1}{4}, \quad c_9 = -\frac{2m_e^*}{\hbar^2} E_{nl}, \\
& c_{10} = 1 + 2\sqrt{\frac{2m_e^*}{\hbar^2}\left(\frac{\ell(\ell+1)\hbar^2}{2m_e^*} + V_0 R_0^2\right) + \frac{1}{4}}, \quad c_{11} = 2\sqrt{-\frac{2m_e^*}{\hbar^2} E_{nl}}, \\
& c_{12} = -\frac{1}{2} + \sqrt{\frac{2m_e^*}{\hbar^2}\left(\frac{\ell(\ell+1)\hbar^2}{2m_e^*} + V_0 R_0^2\right) + \frac{1}{4}}, \quad c_{13} = -\sqrt{-\frac{2m_e^*}{\hbar^2} E_{nl}}.
\end{aligned}
\tag{4.26}
$$

Solving Eq. (4.23) for the parameter values given by Eqs. (4.25)–(4.26), we obtain the energy eigenvalues given in Eq. (4.8). The corresponding eigenfunctions, given in Eq. (4.9), are obtained with the help of Eqs. (4.22) and (4.25)–(4.26), respectively. Note that our results can be recovered as a particular case of the solutions from Refs. [54, 67].

References

1. A. Cho, J. Arthur, Molecular beam epitaxy. Progress in Solid State Chemistry **10**, 157–191 (1975)
2. P.D. Dapkus, Metalorganic chemical vapor deposition. Annual Review of Materials Science **12**(1), 243–269 (1982)
3. K. Watanabe, N. Koguchi, Y. Gotoh, Fabrication of GaAs quantum dots by modified droplet epitaxy. Jpn. J. Appl. Phys. **39**, L79 (2000)
4. A. Tartakovskii, *Quantum Dots: Optics, Electron Transport and Future Applications*. 08 2012
5. F. P. G. de Arquer, D. V. Talapin, V. I. Klimov, Y. Arakawa, M. Bayer, and E. H. Sargent, "Semiconductor quantum dots: Technological progress and future challenges," *Science*, vol. 373, no. 6555, p. eaaz8541, 2021
6. W.A.A. Mohamed, H.A. El-Gawad, S. Mekkey, H. Galal, H. Handal, H. Mousa, A. Labib, Quantum dots synthetization and future prospect applications. Nanotechnol. Rev. **10**(1), 1926–1940 (2021)

7. M.F. Frasco, N. Chaniotakis, Semiconductor quantum dots in chemical sensors and biosensors. Sensors **9**(9), 7266–7286 (2009)
8. D. Heiss, M. Kroutvar, J.J. Finley, G. Abstreiter, Progress towards single spin optoelectronics using quantum dot nanostructures. Solid State Commun. **135**(9), 591–601 (2005)
9. A. Nozik, Quantum dot solar cells. Physica E **14**(1), 115–120 (2002)
10. M.A. Cotta, Quantum dots and their applications: What lies ahead? ACS Appl. Nano Mater. **3**, 4920 (2020)
11. Z. Liu, C.-H. Lin, B.-R. Hyun, Z. Sher, Chin-Wei abd Lv, B. Luo, F. Jiang, T. Wu, C.-H. Ho, H.-C. Kuo, and J.-H. He, "Micro-light-emitting diodes with quantum dots in display technology," *Light Sci. Appl.*, vol. 9, p. 83, 2020
12. S. Jin, Y. Hu, Z. Gu, L. Liu, and H.-C. Wu, "Application of quantum dots in biological imaging," *J. Nanomaterials*, vol. 2011, jan 2011
13. G. Rezaei, Z. Mousazadeh, B. Vaseghi, Nonlinear optical properties of a two-dimensional elliptic quantum dot. Physica E **42**(5), 1477–1481 (2010)
14. G. Rezaei, M. Vahdani, B. Vaseghi, Nonlinear optical properties of a hydrogenic impurity in an ellipsoidal finite potential quantum dot. Curr. Appl. Phys. **11**(2), 176–181 (2011)
15. D. Kaya, M. Ycel, O. Bayrak, The total absorption coefficient and refractive index change of a spherical quantum dot in the confinement radial potential. Physica B **648**, 414403 (2023)
16. W. Xie, A study of an exciton in a quantum dot with Woods-Saxon potential. Superlattices Microstruct. **46**(4), 693–699 (2009)
17. L. Lu, W. Xie, H. Hassanabadi, The effects of intense laser on nonlinear properties of shallow donor impurities in quantum dots with the Woods-Saxon potential. J. Lumin. **131**(12), 2538–2543 (2011)
18. W. Xie, Optical properties of an off-center hydrogenic impurity in a spherical quantum dot with Gaussian potential. Superlattices Microstruct. **48**(2), 239–247 (2010)
19. A. Gharaati, R. Khordad, A new confinement potential in spherical quantum dots: Modified Gaussian potential. Superlattices Microstruct. **48**(3), 276–287 (2010)
20. R. Khordad, Pressure effect on optical properties of modified Gaussian quantum dots. Physica B **407**(7), 1128–1133 (2012)
21. F. Ungan, M. Bahar, M. Barseghyan, L. Prez, D. Laroze, Effect of intense laser and electric fields on nonlinear optical properties of cylindrical quantum dot with Morse potential. Optik **236**, 166621 (2021)
22. R. Khordad, B. Mirhosseini, Linear and nonlinear optical properties in spherical quantum dots: Rosen-Morse potential. Opt. Spectrosc. **117**, 434–440 (2014)
23. F. Ungan, M. K. Bahar, S. Pal, and M. E. Mora-Ramos, "Electron-related nonlinear optical properties of cylindrical quantum dot with the Rosen–Morse axial potential," *Commun. Theor. Phys.*, vol. 72, p. 075505, jun 2020
24. Y. Yakar, B. Çakır, and A. özmen, "Calculation of linear and nonlinear optical absorption coefficients of a spherical quantum dot with parabolic potential," *Opt. Commun.*, vol. 283, no. 9, pp. 1795–1800, 2010
25. B. Çakır, Y. Yakar, A. Özmen, M. Özgür Sezer, and M. Şahin, "Linear and nonlinear optical absorption coefficients and binding energy of a spherical quantum dot," *Superlattices and Microstructures*, vol. 47, no. 4, pp. 556–566, 2010
26. M. Vahdani, The effect of the electronic intersubband transitions of quantum dots on the linear and nonlinear optical properties of dot-matrix system. Superlattices Microstruct. **76**, 326–338 (2014)
27. L. Zhang, X. Li, X. Liu, Z. Li, The influence of second-harmonic generation under the external electric field and magnetic field of parabolic quantum dots. Physica B **618**, 413197 (2021)
28. C.-H. Liu, B.-R. Xu, Theoretical study of the optical absorption and refraction index change in a cylindrical quantum dot. Phys. Lett. A **372**(6), 888–892 (2008)
29. R. Khordad, B. Mirhosseini, Application of Tietz potential to study optical properties of spherical quantum dots. Pramana **85**, 723–737 (2015)
30. D. Hayrapetyan, E. Kazaryan, H. Tevosyan, Optical properties of spherical quantum dot with modified Pöschl-Teller potential. Superlattices Microstruct. **64**, 204–212 (2013)

31. K. Batra, V. Prasad, Spherical quantum dot in Kratzer confining potential: study of linear and nonlinear optical absorption coefficients and refractive index changes. Eur. Phys. J. B **91**, 298 (2018)
32. D.A. Baghdasaryan, E.S. Hakobyan, D.B. Hayrapetyan, H.A. Sarkisyan, E.M. Kazaryan, Nonlinear optical properties of cylindrical quantum dot with Kratzer confining potential. J. Contemp. Phys. **54**, 46 (2019)
33. R. Khordad, Confinement of an exciton in a quantum dot: effect of modified Kratzer potential. Indian Journal of Physics **87**, 623–628 (2013)
34. G. Liu, K. Guo, Linear and nonlinear intersubband optical absorption coefficients and refractive index changes with the ring-shaped non-spherical oscillator potential. Superlattices Microstruct. **52**(5), 997–1009 (2012)
35. M.C. Onyeaju, J.O.A. Idiodi, A.N. Ikot, M. Solaimani, H. Hassanabadi, Linear and nonlinear optical properties in spherical quantum dots: Generalized Hulthén potential. Few-Body Syst. **57**, 793–805 (2016)
36. X. Li, C. Chang, Nonlinear optical properties of $GaAs/Al_\eta Ga_{1-\eta}As$ quantum dots system with Hulthn-Yukawa potential. Opt. Mater. **131**, 112605 (2022)
37. M.C. Onyeaju, J.O.A. Idiodi, A.N. Ikot, M. Solaimani, H. Hassanabadi, Linear and nonlinear optical properties in spherical quantum dots: Manning-Rosen potential. Journal of Optics **46**, 254–264 (2017)
38. Y. Duan, X. Li, Spherical quantum dots in Deng–Fan–Eckart potential confining potential: Study of nonlinear optical rectification. Optik **265**, 169540 (2022)
39. E.B. Al, E. Kasapoglu, H. Sari, I. Skmen, C.A. Duque, Shallow-donor impurity effects on the far infrared electron-electron optical absorption coefficient in single and core/shell spherical quantum dots with Konwent-like confinement potential. Opt. Quant. Electron. **54**, 375 (2022)
40. H. An and J. Motohisa, “Optical properties of InAs quantum dots formed on GaAs pyramids,” *Appl. Phys. Lett.*, vol. 77, pp. 385–387, 07 2000
41. M. Tanaka, H. Yamada, T. Maruyama, K. Akimoto, Well-width dependence of optical properties of rare-earth ion-doped $ZnS_{0.8}Se_{0.2}$/undoped ZnS multiple quantum wells. Phys. Rev. B **67**, 045305 (2003)
42. C.F. Wang, A. Badolato, I. Wilson-Rae, P.M. Petroff, E. Hu, J. Urayama, A. Imamoğlu, Optical properties of single InAs quantum dots in close proximity to surfaces. Appl. Phys. Lett. **85**(16), 3423–3425 (2004)
43. S. Stufler, P. Ester, A. Zrenner, M. Bichler, Quantum optical properties of a single $In_xGa_{1-x}As$-$GaAs$ quantum dot two-level system. Phys. Rev. B **72**, 121301 (2005)
44. X.-L. Wang, V. Voliotis, Epitaxial growth and optical properties of semiconductor quantum wires. J. Appl. Phys. **99**(06), 121301 (2006)
45. E. Lhuillier, S. Keuleyan, P. Guyot-Sionnest, Optical properties of HgTe colloidal quantum dots. Nanotechnology **23**, 175705 (2012)
46. T. Aubert, A.A. Golovatenko, M. Samoli, L. Lermusiaux, T. Zinn, B. Abcassis, A.V. Rodina, Z. Hens, General expression for the size-dependent optical properties of quantum dots. Nano Letters **22**(4), 1778–1785 (2022)
47. D. Ahn and S. lien Chuang, “Calculation of linear and nonlinear intersubband optical absorptions in a quantum well model with an applied electric field,” *IEEE J. Quantum Electron.*, vol. 23, no. 12, pp. 2196–2204, 1987
48. S. Ünlü, I. Karabulut, H. Safak, Linear and nonlinear intersubband optical absorption coefficients and refractive index changes in a quantum box with finite confining potential. Physica E **33**(2), 319–324 (2006)
49. A. F. Nikiforov and V. B. Uvarov, *Special Functions of Mathematical Physics*. Birkhauser Boston, 1988
50. C. Tezcan, R. Sever, A general approach for the exact solution of the Schrödinger equation. Int. J. Theor. Phys. **48**, 337–350 (2009)
51. B.I. Ita, A.I. Ikeuba, Solutions to the Schrödinger equation with inversely quadratic Yukawa plus inversely quadratic Hellmann potential using Nikiforov-Uvarov method. J. Phys. B: At. Mol. **2013**, 1–4 (2013)

52. B.I. Ita, C.O. Ehi-Eromosele, A. Edobor-Osoh, A.I. Ikeuba, Solutions of the Schrödinger equation with inversely quadratic Hellmann plus inversely quadratic potential using Nikiforov-Uvarov method. AIP Conf. Proc. **1629**(1), 360–362 (2014)
53. L. Hitler, B.I. Ita, P.A. Isa, N.-I. Nelson, I. Joseph, O. Ivan, T. Magu, Wkb Solutions for Inversely Quadratic Yukawa plus Inversely Quadratic Hellmann Potential. World J. Appl. Phys. **2**, 109–112 (2017)
54. L. Máthé, C. Onyenegecha, A.-A. FarcaŞ, L.-M. Pioraş-Timbolmaş, M. Solaimani, H. Hassanabadi, Linear and nonlinear optical properties in spherical quantum dots: Inversely quadratic Hellmann potential. Phys. Lett. A **397**, 127262 (2021)
55. C. Berkdemir, *Application of the Nikiforov-Uvarov Method in Quantum Mechanics* (IntechOpen, Rijeka, 2012)
56. D. A. Varshalovich, A. N. Moskalev, and V. K. Khersonskii, *Quantum Theory of Angular Momentum*. WORLD SCIENTIFIC, 1988
57. K.J. Kuhn, G.U. Iyengar, S. Yee, Free carrier induced changes in the absorption and refractive index for intersubband optical transitions in $Al_xGa_{1-x}As/GaAs/Al_xGa_{1-x}As$ quantum wells. J. Appl. Phys. **70**(9), 5010–5017 (1991)
58. R. Khordad, Optical properties of quantum wires: Rashba effect and external magnetic field. J. Lumin. **134**, 201–207 (2013)
59. H. El Ghazi, A. Jorio, I. Zorkani, Linear and nonlinear intra-conduction band optical absorption in (In, Ga)N/GaN spherical QD under hydrostatic pressure. Opt. Commun. **331**, 73–76 (2014)
60. Y. Yakar, B. Çakir, A. Özmen, Linear and nonlinear absorption coefficients of spherical two-electron quantum dot. Comput. Phys. Commun. **188**, 88–93 (2015)
61. I. Karabulut, S. Baskoutas, Linear and nonlinear optical absorption coefficients and refractive index changes in spherical quantum dots: Effects of impurities, electric field, size, and optical intensity. J. Appl. Phys. **103**(7), 073512 (2008)
62. S. Evangelou, Nonlinear optical absorption coefficient and refractive index change in symmetric semiconductor nanostructures: Comparison between different methods. Physica E **124**, 114307 (2020)
63. Y. Duan, X. Li, C. Chang, Z. Zhao, L. Zhang, Hydrostatic pressure, temperature and Al-concentration effects on optical rectification of spherical quantum dots under inversely quadratic Hellmann potential. Optik **254**, 168596 (2022)
64. Y. Duan and X. Li, "The second and third harmonic generation of $GaAs/Ga_{1-x}Al_xAs$ spherical quantum dots under a confining potential," *Eur. Phys. J. B*, vol. 95, May 2022
65. C. Chang, X. Li, Y. Duan, Z. Zhao, Effects of hydrostatic pressure and temperature on the second-harmonic generation of spherical quantum dots with inversely quadratic Hellmann potential. Indian Journal of Physics **97**, 1465–1471 (2023)
66. C. Chang, X. Li, J. Zhuang, Nonlinear third-order optical response of spherical quantum dots with inversely quadratic Hellmann potential. Photonics Nanostructures: Fundam. Appl. **51**, 101029 (2022)
67. C. Chang and X. Li, "Nonlinear optical properties in $gaas/al_{0.3}ga_{0.7}as$ quantum dots of inversely quadratic Hellmann plus Kratzer potential," *Eur. Phys. J. D*, vol. 76, July 2022
68. C. Chang, X. Li, Y. Duan, Z. Zhao, L. Zhang, Optical absorption coefficients of spherical quantum dots system with inversely quadratic Yukawa-Hellmann potential. Physica B **639**, 414009 (2022)
69. Y. Duan, X. Li, C. Chang, Z. Zhao, Calculation of linear and nonlinear optical properties of spherical quantum dots with Hulthn plus Hellmann confining potential. Optik **261**, 169187 (2022)
70. Y.-M. Duan, X.-C. Li, Nonlinear optical rectification of $gaas/ga_{1--x}al_xas$ quantum dots with hulth-en plus hellmann confining potential. Chinese Physics B **32**, 017303 (2023)
71. M. Bansal, V. Kumar, Sahil, S. Bhardwaj, R. M. Singh, and F. Chand, "Investigation of linear and nonlinear optical properties in quantum dots: Influence of magnetic and AB flux fields," *Physica B*, vol. 676, p. 415654, Mar. 2024
72. A. Turkoglu, H. Dakhlaoui, A. S. Durmuslar, M. E. Mora-Ramos, and F. Ungan, "Nonlinear optical properties of a quantum well with inversely quadratic Hellman potential," *Eur. Phys. J. B*, vol. 94, May 2021

Index

© The Editor(s) (if applicable) and The Author(s), under exclusive license to Springer Nature Switzerland AG 2025

L. Máthé, *Physical Properties of Low-Dimensional Systems*,
NanoScience and Technology, https://doi.org/10.1007/978-3-031-91602-1

The manufacturer's authorised representative in the EU is Springer
Nature Customer Service Centre GmbH, Europaplatz 3, 69115 Heidelberg,
Germany. If you have any concerns regarding our products, please
contact ProductSafety@springernature.com

Printed and bound by CPI Group (UK) Ltd, Croydon, CR0 4YY
30/09/2026
02225472-0017